Truncated and Censored Samples from Normal Populations

STATISTICS: Textbooks and Monographs

A SERIES EDITED BY

D. B. OWEN, Coordinating Editor

Department of Statistics
Southern Methodist University
Dallas, Texas

Vol. 1: The Generalized Jackknife Statistic, *H. L. Gray and W. R. Schucany*
Vol. 2: Multivariate Analysis, *Anant M. Kshirsagar*
Vol. 3: Statistics and Society, *Walter T. Federer*
Vol. 4: Multivariate Analysis: A Selected and Abstracted Bibliography, 1957-1972, *Kocherlakota Subrahmaniam and Kathleen Subrahmaniam* (out of print)
Vol. 5: Design of Experiments: A Realistic Approach, *Virgil L. Anderson and Robert A. McLean*
Vol. 6: Statistical and Mathematical Aspects of Pollution Problems, *John W. Pratt*
Vol. 7: Introduction to Probability and Statistics (in two parts), Part I: Probability; Part II: Statistics, *Narayan C. Giri*
Vol. 8: Statistical Theory of the Analysis of Experimental Designs, *J. Ogawa*
Vol. 9: Statistical Techniques in Simulation (in two parts), *Jack P. C. Kleijnen*
Vol. 10: Data Quality Control and Editing, *Joseph I. Naus* (out of print)
Vol. 11: Cost of Living Index Numbers: Practice, Precision, and Theory, *Kali S. Banerjee*
Vol. 12: Weighing Designs: For Chemistry, Medicine, Economics, Operations Research, Statistics, *Kali S. Banerjee*
Vol. 13: The Search for Oil: Some Statistical Methods and Techniques, *edited by D. B. Owen*
Vol. 14: Sample Size Choice: Charts for Experiments with Linear Models, *Robert E. Odeh and Martin Fox*
Vol. 15: Statistical Methods for Engineers and Scientists, *Robert M. Bethea, Benjamin S. Duran, and Thomas L. Boullion*
Vol. 16: Statistical Quality Control Methods, *Irving W. Burr*
Vol. 17: On the History of Statistics and Probability, *edited by D. B. Owen*
Vol. 18: Econometrics, *Peter Schmidt*
Vol. 19: Sufficient Statistics: Selected Contributions, *Vasant S. Huzurbazar (edited by Anant M. Kshirsagar)*
Vol. 20: Handbook of Statistical Distributions, *Jagdish K. Patel, C. H. Kapadia, and D. B. Owen*
Vol. 21: Case Studies in Sample Design, *A. C. Rosander*
Vol. 22: Pocket Book of Statistical Tables, *compiled by R. E. Odeh, D. B. Owen, Z. W. Birnbaum, and L. Fisher*
Vol. 23: The Information in Contingency Tables, *D. V. Gokhale and Solomon Kullback*
Vol. 24: Statistical Analysis of Reliability and Life-Testing Models: Theory and Methods, *Lee J. Bain*
Vol. 25: Elementary Statistical Quality Control, *Irving W. Burr*
Vol. 26: An Introduction to Probability and Statistics Using BASIC, *Richard A. Groeneveld*
Vol. 27: Basic Applied Statistics, *B. L. Raktoe and J. J. Hubert*
Vol. 28: A Primer in Probability, *Kathleen Subrahmaniam*
Vol. 29: Random Processes: A First Look, *R. Syski*

Vol. 30: Regression Methods: A Tool for Data Analysis, *Rudolf J. Freund and Paul D. Minton*

Vol. 31: Randomization Tests, *Eugene S. Edgington*

Vol. 32: Tables for Normal Tolerance Limits, Sampling Plans, and Screening, *Robert E. Odeh and D. B. Owen*

Vol. 33: Statistical Computing, *William J. Kennedy, Jr. and James E. Gentle*

Vol. 34: Regression Analysis and Its Application: A Data-Oriented Approach, *Richard F. Gunst and Robert L. Mason*

Vol. 35: Scientific Strategies to Save Your Life, *I. D. J. Bross*

Vol. 36: Statistics in the Pharmaceutical Industry, *edited by C. Ralph Buncher and Jia-Yeong Tsay*

Vol. 37: Sampling from a Finite Population, *J. Hajek*

Vol. 38: Statistical Modeling Techniques, *S. S. Shapiro*

Vol. 39: Statistical Theory and Inference in Research, *T. A. Bancroft and C.-P. Han*

Vol. 40: Handbook of the Normal Distribution, *Jagdish K. Patel and Campbell B. Read*

Vol. 41: Recent Advances in Regression Methods, *Hrishikesh D. Vinod and Aman Ullah*

Vol. 42: Acceptance Sampling in Quality Control, *Edward G. Schilling*

Vol. 43: The Randomized Clinical Trial and Therapeutic Decisions, *edited by Niels Tygstrup, John M. Lachin, and Erik Juhl*

Vol. 44: Regression Analysis of Survival Data in Cancer Chemotherapy, *Walter H. Carter, Jr., Galen L. Wampler, and Donald M. Stablein*

Vol. 45: A Course in Linear Models, *Anant M. Kshirsagar*

Vol. 46: Clinical Trials: Issues and Approaches, *edited by Stanley H. Shapiro and Thomas H. Louis*

Vol. 47: Statistical Analysis of DNA Sequence Data, *edited by B. S. Weir*

Vol. 48: Nonlinear Regression Modeling: A Unified Practical Approach, *David A. Ratkowsky*

Vol. 49: Attribute Sampling Plans, Tables of Tests and Confidence Limits for Proportions, *Robert E. Odeh and D. B. Owen*

Vol. 50: Experimental Design, Statistical Models, and Genetic Statistics, *edited by Klaus Hinkelmann*

Vol. 51: Statistical Methods for Cancer Studies, *edited by Richard G. Cornell*

Vol. 52: Practical Statistical Sampling for Auditors, *Arthur J. Wilburn*

Vol. 53: Statistical Signal Processing, *edited by Edward J. Wegman and James G. Smith*

Vol. 54: Self-Organizing Methods in Modeling: GMDH Type Algorithms, *edited by Stanley J. Farlow*

Vol. 55: Applied Factorial and Fractional Designs, *Robert A. McLean and Virgil L. Anderson*

Vol. 56: Design of Experiments: Ranking and Selection, *edited by Thomas J. Santner and Ajit C. Tamhane*

Vol. 57: Statistical Methods for Engineers and Scientists. Second Edition, Revised and Expanded, *Robert M. Bethea, Benjamin S. Duran, and Thomas L. Boullion*

Vol. 58: Ensemble Modeling: Inference from Small-Scale Properties to Large-Scale Systems, *Alan E. Gelfand and Crayton C. Walker*

Vol. 59: Computer Modeling for Business and Industry, *Bruce L. Bowerman and Richard T. O'Connell*

Vol. 60: Bayesian Analysis of Linear Models, *Lyle D. Broemeling*

Vol. 61: Methodological Issues for Health Care Surveys, *Brenda Cox and Steven Cohen*

Vol. 62: Applied Regression Analysis and Experimental Design, *Richard J. Brook and Gregory C. Arnold*

Vol. 63: Statpal: A Statistical Package for Microcomputers – PC-DOS Version for the IBM PC and Compatibles, *Bruce J. Chalmer and David G. Whitmore*

Vol. 64: Statpal: A Statistical Package for Microcomputers – Apple Version for the II, II+, and IIe, *David G. Whitmore and Bruce J. Chalmer*

Vol. 65: Nonparametric Statistical Inference, Second Edition, Revised and Expanded, *Jean Dickinson Gibbons*

Vol. 66: Design and Analysis of Experiments, *Roger G. Petersen*

Vol. 67: Statistical Methods for Pharmaceutical Research Planning, *Sten W. Bergman and John C. Gittins*

Vol. 68: Goodness-of-Fit Techniques, *edited by Ralph B. D'Agostino and Michael A. Stephens*

Vol. 69: Statistical Methods in Discrimination Litigation, *edited by D. H. Kaye and Mikel Aickin*

Vol. 70: Truncated and Censored Samples from Normal Populations, *Helmut Schneider*

Vol. 71: Robust Inference, *M. L. Tiku, W. Y. Tan, and N. Balakrishnan*

Vol. 72: Statistical Image Processing and Graphics, *edited by Edward J. Wegman and Douglas J. DePriest*

Vol. 73: Assignment Methods in Combinatorial Data Analysis, *Lawrence Hubert*

Vol. 74: Understanding Statistics, *Bruce J. Chalmer*

Vol. 75: Econometrics and Structural Change, *Lyle D. Broemeling and Hiroki Tsurumi*

Vol. 76: Statistical Tools for Simulation Practitioners, *Jack P. C. Kleijnen*

Vol. 77: Randomization Tests, Second Edition, *E. Edgington*

OTHER VOLUMES IN PREPARATION

Truncated and Censored Samples from Normal Populations

HELMUT SCHNEIDER
Department of Qualitative Business Analysis
Louisiana State University
Baton Rouge, Louisiana

MARCEL DEKKER, INC.   New York and Basel

Library of Congress Cataloging-in-Publication Data

Schneider, Helmut
Truncated and censored samples from normal populations.

(Statistics, textbooks and monographs ; vol. 70)
Includes bibliographies and index.
1. Sampling (Mathematics) 2. Estimation theory.
I. Title. II. Series: Statistics, textbooks and monographs ; v. 70.
QA276.6.S387 1986 519.5'2 86-9031
ISBN 0-8247-7591-0

MARCEL DEKKER, INC.
270 Madison Avenue, New York, New York 10016

Current printing (last digit):
10 9 8 7 6 5 4 3 2 1

PRINTED IN THE UNITED STATES OF AMERICA

To my wife, Mary

Preface

In the last 30 years there has been a significant number of articles in the field of censored samples. Many of these contributions deal with normally distributed data. The normal distribution still plays a significant role in practical situations. In cases where the normal assumption does not hold, however, normality can often be achieved by suitable transformations.

This book contains a collection of results relating to censored samples taken from a normal distribution. The first objective of this monograph has been to assemble scattered materials to help assist applied researchers in solving problems related to censored data taken from normal distributions. Examples are presented from various fields of application such as quality control, life testing, biometrics, economy, and agronomy. The book presents the existing material and gives a comprehensive comparison of known methods of estimation and inference for truncated and censored data from a normal distribution. In several instances of application, gaps are closed; in other instances, problems in application are pointed out.

A secondary aim is to provide materials and references for further research in methodology. Since the objective of the author has been to present results that will be useful to applied researchers in statistics, proofs are not covered.

Chapters 2 and 3 cover the truncated normal distribution, which is closely related to censored normal samples. The remaining chapters deal with the censored samples themselves. Chapter 2 gives some basic properties of the truncated normal distribution which are used throughout the book. Chapter 3 deals with estimation in the truncated normal model.

The remaining part, which deals with censored samples, covers estimation (Chapter 4), linear regression with censored data (Chapter 5), and inference with censored data (Chapter 6). Chapter 7 discusses some applications in quality control and life testing.

Not all of the tables related to the subjects in this book are included; there are excellent tables available elsewhere. References for such tables are given. In citing the references, I have tried to give historical credit where it is due.

This monograph was completed during a visit to the Statistics Department at the University of North Carolina at Chapel Hill. I would like to express my gratitude to Normal L. Johnson for making my visit possible and for contributing valuable suggestions pertaining to my work. I am also indebted to Peter Wilrich for his constant support during the preparation of this manuscript. I extend my appreciation as well for the interest shown to me by Raymond Carroll and other members of the Department of Statistics at the University of North Carolina at Chapel Hill with whom I have had many fruitful discussions. Furthermore, I am grateful to Zelpha Johnson for her assistance in preparing the nomographs. And finally, I would like to thank Ruth Bahr for having the patience and perseverance to handle ably all the typing of a difficult manuscript.

Helmut Schneider

Contents

Truncated and Censored Samples from Normal Populations

CHAPTER 1

Introduction

1.1 GENESIS AND HISTORICAL REMARKS

Truncation of a distribution occurs in many situations. Consider, for instance, the production of an item, the width of which is normally distributed with unknown mean and standard deviation. To confirm that the width of the item is not below a certain limit x_ℓ and not above x_r, all items are checked and sorted. From the screened population (which is now truncated normal) a sample is drawn to estimate the parameters of the original normal distribution.

Detection limits are another field in which the truncated normal distribution is suitable. Quite often instruments measuring data from a normal population have detection limits; i.e., small values $x \leq x_\ell$ and/or large values $x \geq x_r$ are not observable, and therefore their existence is not even reported by the instrument.

Usually it is assumed that the truncation limits are known. If they are unknown estimation can become rather difficult even for light truncation. This is because the bias and variance of the estimators for finite samples turn out to be extraordinarily high.

Single truncation means that $x_\ell = -\infty$ or $x_r = \infty$, while the case $x_\ell > -\infty$ and $x_r < \infty$ is denoted as double truncation. Multiple truncation refers to the situation where values within certain intervals are excluded from observation. A second kind of truncation appears if the limits are unknown but the proportion truncated from the population is known. In literature only the first kind of truncation is considered due to its practical importance.

The term "censoring" is well known in statistical literature and refers to the situation in which the value of some of the observations of the sample is unknown. Consider, for instance, a life test, where the experimenter decides to stop the experiment before all of the units on test have failed. In this case some of the lifetimes are not known, i.e., they are censored. For a censored sample it is assumed that the size of the sample and the number of censored observations are known. If only the number of uncensored observations is known to the experimenter, i.e., the total sample size is unknown, as it is in the case of detection limits, then this kind of censoring is identical with truncation of the population distribution. Some authors use the term "censoring" for this kind of truncation, and some authors use the term "truncation" for censoring a sample. Throughout this book, however, we shall use the term truncation to refer to the restriction of the population prior to sampling. This includes the case of censoring when the sample size is unknown.

The terms "Type I" and "Type II" censoring are used in literature to distinguish between the two different ways of censoring which were also mentioned for the truncated case. Consider again, for instance, a life test. The experimenter might stop the experiment after a preassigned time. This type of censoring is denoted as Type I censoring. In general, Type I censoring refers to the situation where a sample of size n is drawn, and the observations which have values above a predetermined limit or below a predetermined limit are censored. Note that the limits are fixed for Type I censoring while the number of censored observations is random. In life testing the term "time censoring" is more common than Type I censoring.

If the experimenter decides to stop an experiment after a preassigned number of units have failed, then this is called Type II censoring. In general, if the preassigned n_ℓ smallest or n_r largest observations are censored then one refers to Type II censoring. Notice that for Type II censoring the number of censored observations is fixed while the censoring limits can be considered random. In life testing the term "failure censoring" is preferred to Type II censoring. "Multiple censoring" means that observations within certain intervals (Type I) or observations within certain sample percentiles (Type II) are excluded from measurement.

Two other kinds of censoring, namely "arbitrary" (see Nelson, 1982) and "random" censoring (see Kalbfleisch and Prentice, 1980) are not discussed in this book, although the methods presented here can be applied as approximations to these cases. Random censoring refers to the situation in which observations are randomly censored.

Let X_i be the normal random variable and let C_i be another random variable, not necessarily normal; then we observe $Z_i = \min(X_i, C_i)$ and δ_i which is equal to zero when $Z_i = C_i$ (i.e., X_i is censored), and equal to one when $Z_i = X_i$ (i.e., X_i is uncensored). Random censoring appears frequently in clinical trials while Type I and Type II censoring is mainly used in life tests. "Arbitrary censoring" denotes the case where neither Type I nor Type II nor random censoring is used; instead, another arbitrary procedure is used to censor observations.

We shall mention now some of the early works on truncation and censoring for the normal distribution. We shall refer to the newer literature at appropriate places in the text. Perhaps the first writer to consider a sample from a truncated normal distribution was Galton (1898). His objective was to test the suitability of trotting records, provided by the Wallace Year Book and published by the American Trotting Association, for investigations into the laws of heredity. All trotting performances up to the 2 min 30 sec per mile standard are registered in the volumes of Wallace's Year Book. A horse fails to qualify if its speed was less than 2 min 30 sec per mile. The frequency distribution of all recorded speeds per year were thus truncated from below. Galton (1898) showed that a truncated normal distribution would be a suitable description of the data.

A more theoretical analysis of the truncated normal distribution and a derivation of estimators, based on samples from a truncated normal distribution, were given by Pearson and Lee (1908) and Fisher (1931, 1936).

Stevens (1937) considered the estimation of the mean and standard deviation from truncated samples. He applied the results to the truncated time-mortality curve. The term "censoring" for this kind of truncation was, according to Hald (1949), suggested by J. E. Kerrich. Hald (1949) used this term to distinguish the cases where the sample is truncated, from the cases where the distribution is truncated. Although some early papers used the term truncation for censoring a sample, the term "censoring" is now well established in statistical literature and refers to the situation in which the values of some of the observations of a sample are not known.

1.2 APPLICATIONS

Censoring seems to have a much larger field of application than truncation. Truncation appears in the case of detection limits, when the number of censored observations remains unknown, and in quality control problems. Modern technology often allows a production to be automatically screened. As a result of such screening procedures, a formerly normal population becomes truncated.

Censoring is most often used in life tests and clinical trials. Although life times are usually not normally distributed, the logarithm of the times can frequently be treated as normal. There is an important competitor of the lognormal distribution, namely the Weibull distribution. In fact, in many cases a test will hardly be able to discriminate between the lognormal and the Weibull model unless the sample size is sufficiently large.

Tiku (1980) proposes applying Type II symmetrical censoring ($n_\ell = n_r$) to obtain robust estimators for μ and σ. He shows that this method performs quite well in comparison with other well-known non-parametric robust estimators.

We shall now give some practical examples where censoring is applied.

Single Type II Censored Sample in an Experiment

The data in Table 1.1 (Gupta, 1952, p. 271) show the days on which the first seven of a sample of ten tested mice died after being inoculated with a uniform culture of human tuberculosis.

The objective is to estimate the median survival time. Assuming that the log-data are normally distributed, we shall discuss several methods of estimation. We shall also present methods of testing the suitability of the normal model and of drawing inference given the normal model.

Double Type I Censored Sample in a Time-Mortality Experiment

In a time-mortality experiment (Cohen, 1957), the first observation is delayed until the collapse of a fixed time interval, with the result that n_ℓ sample specimens die before observation begins. The experiment is subsequently terminated at a predetermined time with n_r

TABLE 1.1 Survival Time of Inoculated Mice

Days after inoculation	41	44	46	54	55	58	60
Log days after inoculation	1.613	1.643	1.663	1.732	1.740	1.763	1.778

TABLE 1.2 Hours to Failure of Motorettes

150°C	All 10 motorettes without failure at 8064 hours
170°C	1764, 2772, 3444, 3542, 3780, 4860, 5196 3 motorettes without failure at 5448 hours
190°C	408, 408, 1344, 1344, 1440 5 motorettes without failure at 1680 hours
220°C	408, 408, 504, 504, 504 5 motorettes without failure at 528 hours

specimens remaining alive. Actual survival times are recorded for the n_0 specimens which die during the period of full observation. For a specific sample of this type, X, the log survival time in days is assumed to be normally distributed. The objective here is again to estimate median life.

Type I Censoring in an Accelerated Life Test

A set of data which has been the topic of much discussion is due to Crawford (1970), who considered the failure times of electrical insulation of motorettes as a function of temperature. Ten motorettes were tested at each of the temperatures 220°C, 190°C, 170°C, and 150°C, with interest focusing on the median and 10th percentile of the distribution at 130°C, which, as is typical in accelerated life tests, was unobserved. The Arrhenius law, which is a chemical rate equation, was used. This law suggests that the mean logarithmic life as a function of the absolute temperature T can be written as

$$\text{mean log life} = \beta_0 + \beta_1/T$$

In order to demonstrate the effect of censoring on estimation, data were selected at various times during the experiment, i.e., after 15, 16, 19, 21, and 33 months. The example is analyzed in greater detail in Chapter 5. The 16th month data are listed in Table 1.2. The censoring in this example is of Type I. The censoring times are 8064 h, 5448 h, 1680 h, and 528 h, respectively. However, this experiment, in which considerable control is available to the experimenter, also offers the possibility of Type II censoring. For instance, in the above study, it might have been decided in advance to continue the experiment until specified numbers of motorettes had failed at each of the temperatures (e.g., until one, seven, five, and four motorettes had failed at 150°C, 170°C, 190°C, and 220°C, respectively). In some models inference is simpler for Type II censoring than for Type I censoring. We shall come back to this problem in Section 1.3. Further examples for accelerated life tests are given in Nelson (1982).

Random Censoring for the Heart Transplant Data

Miller and Halpern (1982) give survival times of potential heart transplant recipients from their date of acceptance into the Stanford Heart Transplant Program.

One problem of considerable interest is the evaluation of the effect of age at transplantation on survival time. The data given by Miller and Halpern report the situation in February 1980. Some of the patients died; others are still alive; thus, their life time is censored. A patient has to wait until a suitable donor is located and transplantation takes place, at which time he joins the treatment group. Hence, a patient's joining the group occurs randomly; subsequently, the censoring time can also be considered random. In literature (see Miller and Halpern, 1982) several non-parametric methods have been used to estimate the slope and intercept of the model log life $= \beta_0 + \beta_1$ Age $+ \epsilon$. In Chapter 5 we compare the very popular non-parametric method of Buckley and James (1979) with the maximum likelihood estimator based on Type I censoring.

The Engle curve relationship. Tobin (1958) considers as an example the Engle curve relationship, showing how household expenditures on various categories of goods vary with household income. For many categories (luxuries) there are no expenditures for low income levels. A single straight line cannot, therefore, represent the Engle curve for both low and high income. If individual households were identical except for income level, the Engle curve would be a broken line since the expenditure would be zero up to a certain income and increase linearly with income afterwards. In this example the dependent variable of the regression has a lower limit, namely zero.

An example from agronomy. Scott and Johnes (1985) discuss a problem from agronomy. The purpose of their investigation was to quantify the effect of a wide range of suboptimal temperatures (i.e., 6-20°C) on seed germination and to compare responses among several

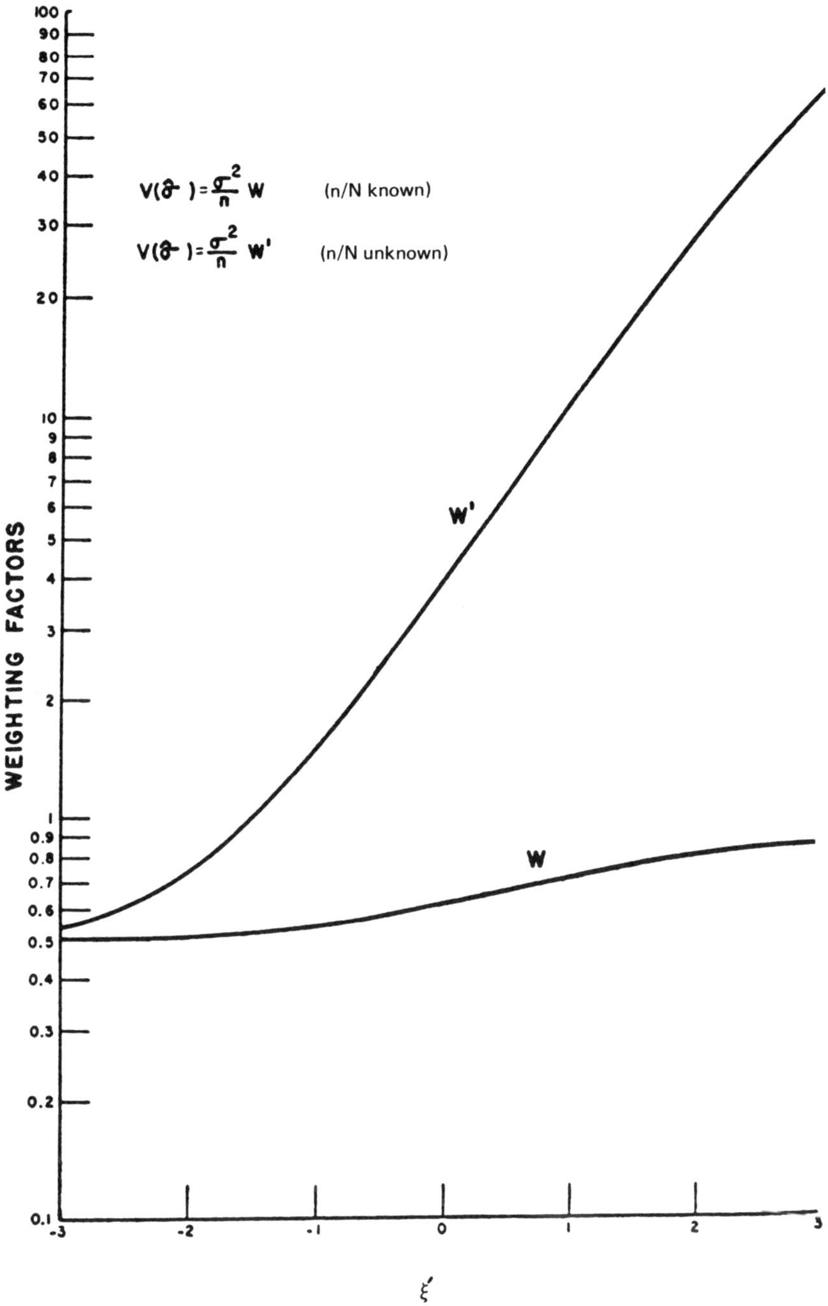

FIGURE 1.1 Weighting factors for use in determining the variance of $\hat{\sigma}$. $\xi' = \hat{u}_\ell = (x_\ell - \hat{\mu})/\hat{\sigma}$; n/N proportion uncensored. (From Cohen, 1950.)

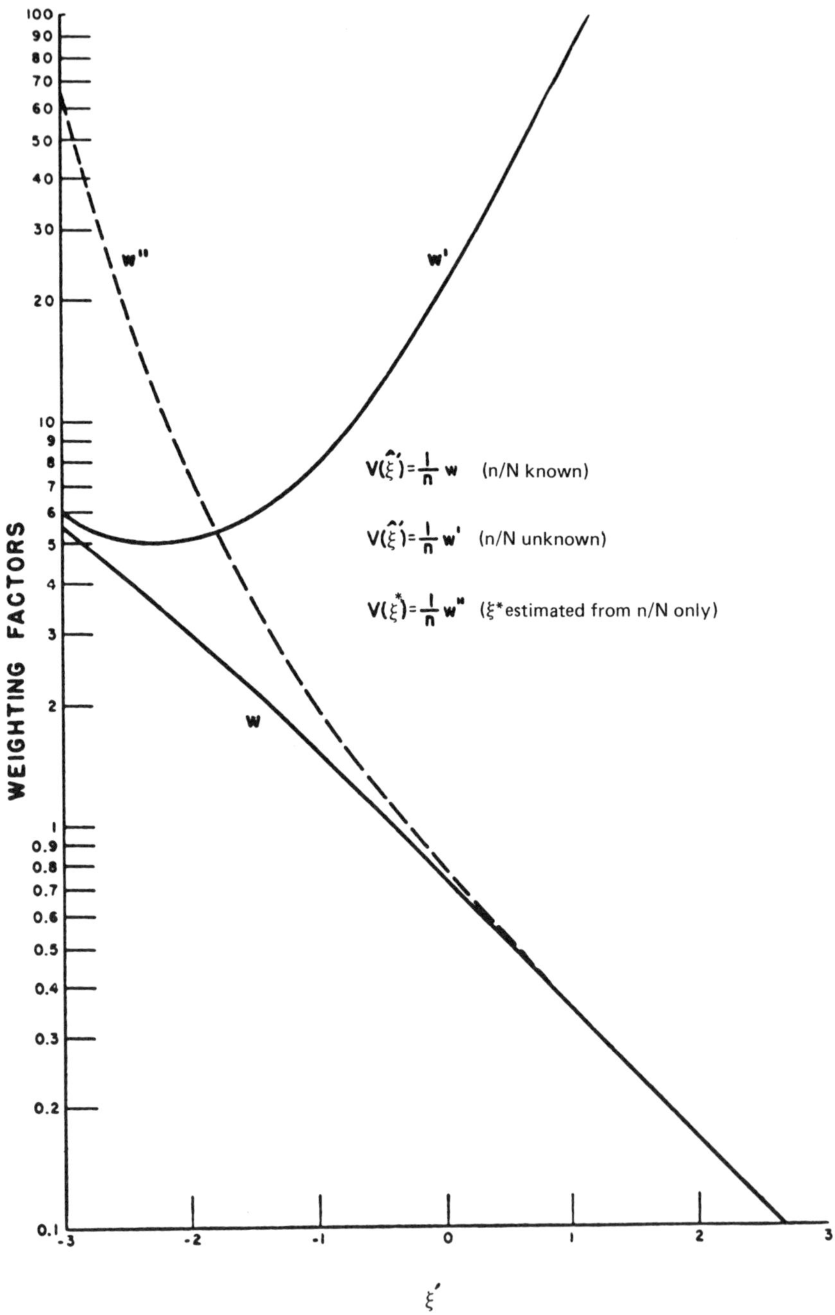

FIGURE 1.2 Weighting factors for use in determining the variances of $\xi' = \hat{u}_\ell = (x_\ell - \hat{\mu})/\hat{\sigma}$; n/N proportion uncensored. (From Cohen, 1950.)

high altitude ecotypes of wild Lycopersicon species and selected cultivated accessions (cultivated tomato). As the low temperature limit for germination is approached, response times increase rapidly. As a result, some viable seeds commonly fail to germinate before an experiment is terminated. Those viable, nonresponding seeds are Type I censored observations.

1.3 VARIANCE OF ESTIMATORS

We shall now make some general remarks about the variance of the estimators. Compared with Type I censoring, random censoring certainly increases the variance of the estimators for the model parameter. Type I censoring also results in a lower variance of the estimator than truncation due to the higher information available.

Figures 1.1 and 1.2, prepared by Cohen (1950), compare the variances of the maximum likelihood estimators $\hat{\mu}$ and $\hat{\sigma}$ in the singly truncated and singly Type I censored case. This comparison serves to indicate the extent to which information contained in a single truncated sample is increased by adding to it our knowledge of the number of unmeasured observations. For a demonstration of the extremely high variance of the estimator $(x_\ell - \hat{\mu})/\hat{\sigma} = \hat{\mu}_\ell$ in the truncated case, a graph is also included in Figure 1.2 of the weighting factor applicable in determining the variance of u_ℓ^*, where u_ℓ^* is only estimated from the proportion of uncensored observations n_0/n.

In principle, the asymptotic covariances of Type I and Type II censoring are identical, given the Type I censored proportions $p_r = 1 - \Phi(u_r)$ and $p_\ell = \Phi(u_\ell)$ are equal to the Type II censored proportions n_r/n and n_ℓ/n, respectively. But there is one major difference. For Type II censoring n_r/n and n_ℓ/n are fixed and so is the covariance matrix up to unknown σ^2. For Type I censoring p_r and p_ℓ are random variables and the covariance matrix has to be estimated. A common method is to replace p_r and p_ℓ by their estimates $\hat{p}_r = 1 - \Phi(\hat{u}_r)$ and $\hat{p}_\ell = \Phi(\hat{u}_\ell)$, respectively. But this fact makes it more complicated to draw inferences with Type I censored samples than with Type II censored samples. More accurate confidence regions can thus be obtained for Type II censoring.

REFERENCES

Buckley, J. and James, I. (1979). Linear regression with censored data, Biometrika 66, 429-436.

Cohen, A. C. (1950). Estimating the mean and variance of normal populations from singly truncated and doubly truncated samples, Annals of Mathematical Statistics 21, 557-569.

Cohen, A. C. (1957). On the solution of estimating equations for truncated and censored samples from normal population, Biometrika 44, 225-236.

Crawford, D. E. (1970). Analysis of incomplete life test data on motorettes, Insulation/Circuits 16, 10, 43-48.

Fisher, R. A. (1931). Properties and applications of H functions, in Mathematical Tables, Vol. 1, XXVI-XXXV, British Association for the Advancement of Science.

Fisher, R. A. (1936). Statistical Methods for Research Workers, 6th ed., Oliver and Boyd.

Galton, F. (1898). An examination into registered speeds of American trotting horses, with remarks on their value as hereditary data, Proc. Royal Soc. of London 62, 310-314.

Gupta, A. K. (1952). Estimation of the mean and standard deviation of a normal population from censored sample, Biometrika 39, 260-273.

Hald, A. (1949). Maximum likelihood estimation of the parameters of a normal distribution which is truncated at a known point, Scandinavian Actuarial Journal 32, 119-134.

Kalbfleisch, J. D. and Prentice, R. L. (1980). The Statistical Analysis of Failure Time Data, New York: Wiley.

Miller, R. G. and Halpern, J. (1982). Regression with censored data, Biometrika 69, 521-531.

Nelson, W. (1982). Applied Life Data Analysis, New York: Wiley.

Pearson, K. and Lee, A. (1908). On the generalized probable error in multiple normal correlation, Biometrika 6, 59-68.

Scott, S. J. and Johnes, R. A. (1985). Quantifying seed germination responses to low temperatures: Variation among Lycopersicon spp., Environmental and Experimental Botany 25, 1-9.

Stevens, W. L. (1937). The truncated normal distribution, Appendix to paper by C. I. Bliss on: The calculation of time mortality curve, Ann. Appl. Biol. 24, 815-852.

Tiku, M. L. (1980). Robustness of MML estimators based on censored samples and robust test statistics, Journal of Statistical Planning and Inference 4, 123-143.

Tobin, J. (1958). Estimation of relationships for limited dependent variables, Econometrica 26, 24-36.

CHAPTER 2

Properties of the Truncated Normal Distribution

This chapter contains some basic results concerning the properties of the truncated normal distribution. It may be used for purposes of reference. Throughout the book specific references are made to results in this chapter.

2.1 DEFINITIONS AND NOTATIONS

Some of the following results appear in Johnson and Kotz (1970a, p. 81, et seq.).

The probability density function (pdf) of a standard normal random variable U is given by

$$\phi(u) = \frac{1}{\sqrt{2\pi}} \exp\left[-\frac{u^2}{2}\right] \qquad -\infty \leq u \leq \infty \tag{2.1}$$

Let $\Phi(u)$ denote the cumulative distribution function (cdf) of U. The notations ϕ and Φ are used throughout the book only in reference to normal distributions.

We shall denote the quantiles of U by u_α, where $u_\alpha = -u_{1-\alpha}$ and

$$\Pr(U \leq u_{1-\alpha}) = \Phi(u_{1-\alpha}) = 1 - \alpha$$

The cdf of a normal random variable X with mean μ and variance ^{2} is $\Phi[(x - \mu)/\sigma]$. The random variables X and U are linearly related and the relationship is given by $U = (X - \mu)/\sigma$.

A random variable X has a doubly truncated normal distribution if its probability density function is

$$f(x;\mu,\sigma,x_\ell,x_r) = \begin{cases} 0 & \text{when} \quad x < x_\ell \\ \dfrac{1}{\sqrt{2\pi}\,\sigma} \exp\left[-\dfrac{(x-\mu)^2}{2\sigma^2}\right] \times \left\{\dfrac{1}{\sqrt{2\pi}\,\sigma}\displaystyle\int_{x_\ell}^{x_r} \exp\left[-\dfrac{(t-\mu)^2}{2\sigma^2}\right] dt\right\}^{-1} & \text{when} \quad x_\ell \leqslant x \leqslant x_r \\ 0 & \text{when} \quad x > x_r \end{cases}$$

where x_ℓ and x_r are the left (lower) and right (upper) truncation points, respectively. Using the notation

$$u_r = (x_r - \mu)/\sigma \tag{2.2}$$

$$u_\ell = (x_\ell - \mu)/\sigma \tag{2.3}$$

the pdf of a truncated normal distribution can be written as

$$f(x;\mu,\sigma,x_\ell,x_r) = \begin{cases} \dfrac{\phi[(x-\mu)/\sigma]}{\sigma[\Phi(u_r) - \Phi(u_\ell)]} & u_\ell \leqslant \dfrac{x-\mu}{\sigma} \leqslant u_r \\ 0 & \text{otherwise} \end{cases} \tag{2.4}$$

The degrees of truncation are $\Phi(u_\ell)$ on the left and $1 - \Phi(u_r)$ on the right. If x_ℓ is replaced by $-\infty$, or x_r by ∞, the distribution is singly truncated from the right, or the left, respectively.

The following expressions are extensively used in the later chapters. For a doubly truncated normal distribution we define

$$Z_r(u_\ell, u_r) = \phi(u_r)/[\Phi(u_r) - \Phi(u_\ell)] \tag{2.5}$$

$$Z_\ell(u_\ell, u_r) = \phi(u_\ell)/[\Phi(u_r) - \Phi(u_\ell)] \tag{2.6}$$

For a singly truncated normal distribution we define

$$W(u_r) = Z_r(-\infty, u_r) = \phi(u_r)/\Phi(u_r) \tag{2.7}$$

$$\tilde{W}(u_\ell) = Z_\ell(u_\ell, \infty) = \phi(u_\ell)/[1 - \Phi(u_\ell)] \tag{2.8}$$

Since $\tilde{W}(u_\ell) = W(-u_\ell)$, we might use W(u) also in the left truncated case. Note that the reciprocal of $\tilde{W}(u)$ is called Mills' ratio, because of its tabulation by Mills (1926). Tables of $1/\tilde{W}(u)$ which are important in connection with the subject discussed here are also included in Pearson (1955) and Sheppard (1939). A computer program for calculating $\tilde{W}(u)$ is given by Swan (1969c). Several approximations to Mills' ratio have been suggested in literature (see Johnson and Kotz, 1970b). For a comprehensive discussion of Mills' ratio we refer to Patel and Read (1982).

More complicated forms of truncation, e.g., omission of a number of intervals of X, are described by the pdf

$$f(x; \mu, \sigma, x_{(i)}, x^{(i)}) = \begin{cases} \dfrac{\phi[(x-\mu)/\sigma]}{\sigma\Sigma\{\Phi[(x^{(i)}-\mu)/\sigma] - \Phi[(x_{(i)}-\mu)/\sigma]\}} & x \in [x_{(i)}, x^{(i)}] \\ 0 \quad \text{when} & x \in [x^{(i)}, x_{(i+1)}] \\ i = 1, 2, \ldots, \ell \end{cases} \tag{2.9}$$

2.2 THE MOMENTS

The moments of a truncated normal distribution can be expressed in terms of the I_n functions (Fisher, 1931, and Cohen and Woodward, 1953) defined by

$$\begin{aligned} I_n(u) &= \frac{1}{\sqrt{2\pi}} \int_0^\infty \frac{t^n}{n!} \exp[-(t+u)^2/2]\, dt \qquad n \geq 1 \\ I_0(u) &= 1 - \Phi(u) \\ I_{-1}(u) &= \phi(u) \end{aligned} \tag{2.10}$$

$I_n(u)$ satisfies the recurrence equation

$$(n+1)I_{n+1}(u) + uI_n(u) - I_{n-1}(n) = 0 \qquad n \geq 0 \tag{2.11}$$

Let

$$Y = X - x_\ell$$

and

$$d = x_r - x_\ell \tag{2.12}$$

then the pdf of Y is

$$f(y;\mu,\sigma,u_r,u_\ell) = \begin{cases} \dfrac{\phi(y/\sigma + u_\ell)}{\sigma[\Phi(u_r) - \Phi(u_\ell)]} & 0 \leq y \leq d \\ 0 & \text{otherwise} \end{cases} \tag{2.13}$$

We calculate the moments of Y by

$$m_k = \frac{\int_0 y^k \phi(y/\sigma + u_\ell)\, dy}{\sigma\,[\Phi(u_r) - \Phi(u_\ell)]} \qquad k \geq 1 \tag{2.14}$$

These moments can now easily be expressed in terms of I_n

$$m_k = \frac{k!\sigma^k \left[I_k(u_\ell) - \sum_{i=0}^{k} (d/\sigma)^{k-i} I_i(u_r)/(k-i)! \right]}{I_0(u_\ell) - I_0(u_r)} \tag{2.15}$$

For m_1 we obtain

$$m_1 = \sigma \frac{I_1(u_\ell) - I_1(u_r)}{I_0(u_\ell) - I_0(u_r)} - d \frac{I_0(u_r)}{I_0(u_\ell) - I_0(u_r)}$$

or, using the notations Z_ℓ and Z_r,

$$m_1 = \sigma[Z_\ell - Z_r - u_\ell] \tag{2.16}$$

where, for the sake of convenience, the arguments u_ℓ and u_r have been omitted. The moments m_k for $k \geq 2$ can be obtained from the recurrence equation

$$m_k = -\sigma Z_r d^{k-1} + (k-1)\sigma^2 m_{k-2} - \sigma u_\ell m_{k-1} \tag{2.17}$$

where $m_0 = 1$. This equation is derived from (2.11) and (2.15) by a straightforward evaluation. The second moment of Y is thus given by

$$m_2 = -\sigma Z_r d + \sigma^2 - \sigma\mu_\ell\, m_1$$

or, substituting m_1, by the right-hand side of (2.16)

$$m_2 = \sigma^2[1 - Z_r(u_r - u_\ell) - u_\ell(Z_\ell - Z_r) + u_\ell^2]$$

The expected value and variance of X are thus

$$E[X] = \mu + \sigma(Z_\ell - Z_r) \tag{2.18}$$

$$V[X] = \sigma^2[1 + u_\ell Z_\ell - u_r Z_r - (Z_\ell - Z_r)^2] \tag{2.19}$$

If the normal distribution is singly truncated on the left we obtain

$$E[X] = \mu + \quad W(-u_\ell) \tag{2.20}$$

$$V[X] = \sigma^2\{1 - W(-u_\ell)[W(-u_\ell) - u_\ell]\} \tag{2.21}$$

$$m_k = \sigma^2(k-1)m_{k-2} - \sigma u_\ell m_{k-1} \qquad k \geq 2,\ m_0 = 1 \tag{2.22}$$

If the normal distribution is singly truncated on the right, we substitute $-u_\ell$ by u_r. Hence

$$E[X] = \mu - \sigma W(u_r) \tag{2.23}$$

$$V[X] = \sigma^2 \{1 - W(u_r)[W(u_r) + u_r]\} \tag{2.24}$$

$$m_k = \sigma^2 (k - 1) m_{k-2} + \sigma u_r m_{k-1} \qquad k \geq 2,\ m_0 = 1 \tag{2.25}$$

Throughout the book we shall use the common notation $\hat{\mu}$ to refer to an estimate of the parameter μ. When we discuss special estimators such as the maximum likelihood estimator we shall use the abbreviation μ_{ML} omitting "^".

2.3 LOG-TRANSFORMATION

In many applications the lognormal distribution is used because it has a great variety of shapes and empirically fits many types of data adequately. The close relationship to the normal distribution, the fact that the logarithm of the data is normally distributed, makes it easy to deal with the lognormal distribution. The cumulative distribution function of the lognormal distribution is

$$F(x) = \Phi \{[\log_{10}(y) - \mu]/\sigma\}, \qquad y > 0 \tag{2.26}$$

In most of the applications the common (base 10) logarithm is used. Throughout the book $\log(\cdot)$ is used for the natural (base e) logarithm, while $\log_{10}(\cdot)$ is used for the common logarithm. We also note that throughout antilog(x) is used for 10^x.

The parameters of the lognormal distribution are μ and σ. Some of the important characteristics are given below. Note that the common logarithm is used.

100pth Lognormal percentile:

$$y_p = \text{antilog}(\mu + u_p \sigma) \tag{2.27}$$

Lognormal median:

$$y_{0.5} = \text{antilog}(\mu) \tag{2.28}$$

Lognormal mean:

$$E[Y] = \text{antilog}[\mu + 0.5 \log(10) \sigma^2] \tag{2.29}$$

Lognormal variance:

$$\text{var}[Y] = E[Y]^2\{\text{antilog}[\log(10)\sigma^2] - 1]\} \tag{2.30}$$

Further characteristics are given by Johnson and Kotz (1970a).

REFERENCES

Cohen, A. C. and Woodward, J. (1953). Tables of Pearson-Lee-Fisher functions of singly truncated normal distributions, Biometrics 9, 489-497.

Fisher, R. A. (1931). Properties and applications of H functions, in Mathematical Tables, Vol. 1, XXVI-XXXV, British Association for the Advancement of Science.

Johnson, N. L. and Kotz, S. (1970a). Continuous Distributions, Vol. 1, New York: Houghton-Mifflin.

Johnson, N. L. and Kotz, S. (1970b). Continuous Univariate Distributions, Vol. 2, New York: Houghton-Mifflin.

Lee, A. (1915). Table of Gaussian tail function when the tail is larger than the body, Biometrika 10, 208-215.

Mills, J. P. (1926). Table of the ratio: Area to bounding ordinate for any portion of normal curve, Biometrika 18, 295-400.

Patel, J. K. and Read, C. B. (1982). Handbook of the Normal Distribution, New York: Marcel Dekker.

Pearson, E. S. (1955). The normal probability function: Tables of certain area-ordinate ratios and of their reciprocals, Biometrika 42, 217-222.

Sheppard, W. F. (1939). The probability integral, in British Association Mathematical Tables, Vol. 7, London: Cambridge University Press.

Swan, A. V. (1969). Algorithm AS17. The reciprocal of Mill's ratio, Applied Statistics 18, 115-116.

CHAPTER 3

Estimators for the Parameters of a Truncated Normal Distribution

Truncation of a normal distribution usually refers to a truncation at a certain point or points. The truncation points might be known or unknown. Another kind of truncation can also be considered where a known proportion is truncated from the population. Before proceeding in detail, we shall briefly discuss examples in which truncations might appear.

Let us consider a production process of pills the weights of which have to lie within certain limits. The weight can usually be assumed to be normally distributed. An automatic screening procedure might be used to sort out the pills. Suppose they are sorted out according to their weight; then, as a result of the screening procedure, we have a truncated normal distribution which is truncated at certain points. Often it is the case that other quality characteristics such as pressure are used, during the production process to sort out the pills. In this case the true truncation points are not known and have to be estimated. But since the number of sorted pills might be recorded at the end of a production cycle, we have the true proportion truncated from the entire population.

This chapter will begin with the truncation at a certain known point. In Section 3.5 the case of unknown truncation points is discussed and in Section 3.6 we consider the truncation of a known proportion. The solution to any of the estimation problems may be obtained by the maximum likelihood (ML) method. The method of moments considered by Pearson and Lee (1908) yields the same estimators as first pointed out by Fisher (1931).

Since the likelihood equations are usually solvable only by iterative methods, moments (MM) estimators were suggested by Cohen (1950b) and Shah and Jaiswal (1966) which give explicit results. These estimators use higher moments and thus have the advantage of avoiding iterative procedures but have the disadvantage of loss of efficiency. For a comparison of the ML-estimators and the MM-estimators we refer to Section 3.3.

A third approach is to seek the linear function of the available order statistics, which is unbiased with minimum variance (MV), in estimating the parameters μ and σ. The method of least squares (LS) is applied to the ordered observations. To do this the expectation vector and dispersion matrix of the order-statistics have to be calculated for the truncated distribution. But since the dispersion matrix is a function of the truncation points x_ℓ and x_r, the MV unbiased linear function will differ as x_ℓ and x_r vary. No work has been done yet concerning this approach. We will not consider it in this book.

Section 3.3. points out some problems in estimating the parameters μ and σ when truncation becomes severe.

3.1 MAXIMUM LIKELIHOOD ESTIMATORS

Maximum likelihood estimators were considered by Hald (1949) and Cohen (1949, 1950a, 1957, 1959, 1961). We will first consider the singly truncated case.

The Singly Truncated Normal Distribution

Let X be a normally distributed random variable truncated at the right point x_r. The pdf is

$$f(x; \mu, \sigma, x_r) = \begin{cases} \dfrac{\phi[(x - \mu)\sigma]}{\sigma\Phi[(x_r - \mu)/\sigma]} & x \leq x_r \\ 0 & x > x_r \end{cases} \tag{3.1}$$

Because of the symmetry of the normal distribution a left truncated normal random variable does not have to be considered separately.

Suppose a sample $(x_1, x_2, \ldots, x_n)$ of size n is given from a population with pdf $f(x; \mu, \sigma, x_r)$. With $u_r = (x_r - \mu)/\sigma$ the likelihood function of the sample is

$$L(\mu, \sigma) = [\sigma\sqrt{2\pi}\,\Phi(u_r)]^{-n} \exp\left\{\sum_{i=1}^{n} -[(x_i - \mu)^2]/2\sigma^2\right\}$$

The natural logarithm of the likelihood function is given up to a constant

$$\log L(\mu, \sigma) = -n \log \Phi(u_r) - n \log \sigma - 1/(2\sigma^2) \sum_{i=1}^{n} (x_i - \mu)^2$$

If both μ and σ are unknown the estimates μ_{ML} and σ_{ML} are solutions to the likelihood equations

$$\frac{\partial \log L(\mu, \sigma)}{\partial \mu} = n \frac{\phi(u_r)}{\Phi(u_r)} \frac{1}{\sigma} + \frac{1}{\sigma^2} \Sigma (x_i - \mu) = 0 \tag{3.2}$$

$$\frac{\partial \log L(\mu, \sigma)}{\partial \sigma} = n \frac{\phi(u_r)}{\Phi(u_r)} \frac{u_r}{\sigma} - \frac{n}{\sigma} + \frac{1}{\sigma^3} \Sigma (x_i - \mu)^2 = 0 \tag{3.3}$$

In what follows we use the notation

$$\bar{x} = \frac{1}{n} \sum_{i=1}^{n} x_i \tag{3.4}$$

$$s^2 = \frac{1}{n} \sum_{i=1}^{n} (x_i - \bar{x})^2 \tag{3.5}$$

$$\delta = \frac{\mu_{ML} - \bar{x}}{\sigma_{ML}} \tag{3.6}$$

Cohen (1959) suggested the following simplifications

$$\delta = W(u_r) \tag{3.7}$$

$$s^2 + (\bar{x} - \mu)^2 - \sigma^2 \{1 - u_r[1 - u_r W(u_r)]\} = 0 \tag{3.8}$$

where $W(\cdot)$ was defined by $W(u) = \phi(u)/\Phi(u)$.

σ^2 Known. We define

$$\theta = \frac{x_r - \bar{x}}{\sigma} \tag{3.9}$$

The likelihood equation (3.7) can be written as

$$\delta = W(\theta - \delta) \tag{3.10}$$

and hence the ML-estimator is

$$\mu_{ML} = \bar{x} + \delta\sigma \tag{3.11}$$

where δ is the unique solution to (3.10). Since the likelihood equation (3.10) is solvable only by iterative methods, approximations are presented below. They allow the determination of the value δ, for a given θ, without using an iterative procedure.

Taylor approximation. A Taylor expansion yields

$$\delta = W(\theta) - \delta W'(\theta) + R(\theta) \tag{3.12}$$

where

$$W'(\theta) = -[\phi^2(\theta) + \theta\phi(\theta)\Phi(\theta)]/\Phi^2(\theta) \tag{3.13}$$

and thus the following approximation can be derived

$$\delta = W(\theta)/[1 + W'(\theta)] + \epsilon(\theta)$$

The approximation error $\epsilon(\theta) \rightarrow 0$ as $\theta \rightarrow \infty$. If $\theta \geqslant 2$, the maximum absolute relative error, $|\epsilon(\theta)/\delta|$, is 0.007.

Rational approximation. Several approximations by rational functions were tested. The following approximation was found to be suitable where less than 50% of the population is truncated, i.e., $u_r \geqslant 0$.

$$\delta = P_3(\theta)/P_2(\theta) + \epsilon(\theta)$$

If $0.7979 \leq \theta \leq 2$ then $|\epsilon(\theta)| \leq 1.3 \times 10^{-5}$ for

$$P_3(\theta) = 17.79998379 - 19.306575250\theta + 7.229392410\theta^2 - 0.935797420\theta^3$$

$$P_2(\theta) = 1 + 12.023480020\theta - 3.798744500\theta^2$$

If $2 < \theta \leq 4.3$ then $|\epsilon(\theta)| \leq 2 \times 10^{-5}$ for

$$P_3(\theta) = 0.36123448 - 0.26136921\theta + 0.06349023\theta^2 - 0.00517176\theta^3$$

$$P_2(\theta) = 1 - 0.92775053\theta + 0.41747926\theta^2$$

If $\theta > 4.3$ then $|\epsilon(\theta)| \leq 10^{-5}$ for

$$\delta = 0$$

A nomograph for determining δ in the known σ^2 case is given in Appendix A, Figure A1.

Asymptotic variance. The ML-estimator μ_{ML} is consistent, asymptotically normally distributed and efficient. The asymptotic variance is

$$\text{var}(\mu_{ML}) = \frac{1}{n} \frac{1}{E[(\partial \log f(x, \mu, \sigma, x_r)/\partial\mu)^2]}$$

$$= \frac{\sigma^2}{n} \frac{1}{1 - W(u_r)[W(u_r) + u_r]} \tag{3.14}$$

where $u_r = (x_r - \mu_{ML})/\sigma$.

We have restricted our considerations to the estimation given a right truncated normal distribution. But note that an ML-estimator of μ for a left truncated normal distribution can be obtained by setting $\theta = -(x_\ell - \bar{x})/\sigma$, $\delta = -(\mu_{ML} - \bar{x})/\sigma$ and substituting u_ℓ for u_r. Afterwards we proceed as in the right truncated case.

Example 3.1

To illustrate estimation in the singly right truncated case with known variance, we consider an example in which the entire production of a certain part of a car is sorted through go, no-go gauges. As a result items of diameter in excess of 1.2030 inches are discarded. From experience it is known that the variance of the production is $\sigma^2 = 3.6 \times 10^{-5}$. For a random sample of 75 items selected from the screened production, we obtain

$$\bar{x} = 1.1945 \text{ in.}$$

In order to estimate μ we first calculate $\theta = (1.203 - 1.1945)/0.006 = 1.417$. From Figure A1 of Appendix A we read $\delta(1.417) \simeq 0.2$. The MLE is thus

$$\begin{aligned} \mu_{ML} &= 1.1945 + 0.2 \times 0.006 \\ &= 1.957 \end{aligned}$$

A more accurate estimate is obtained by using the rational approximation. Since $\theta \leqslant 2$, we do not use the Taylor approximation. The first rational function gives

$$\delta = 0.22055$$

and thus the MLE is

$$\mu_{ML} = 1.195823$$

The variance of the estimator is approximately given by (3.14). Thus substituting $\hat{u}_r = 1.195$ in equation (3.14), we obtain

$$\text{var}(\mu_{ML}) \simeq 1.45 \frac{\sigma^2}{n} = 6.98 \times 10^{-7}$$

and the standard error is approximately 8.4×10^{-4}.

Example 3.2

To illustrate estimation in the left truncated case with known variance we consider the same situation as in Example 3.1 but assume that all items of diameter less than 1.186 inches are discarded. Note that we have chosen $x_\ell = \bar{x} - (x_r - \bar{x})$ for the purpose of illustration. Let us further assume $\sigma = 0.006$ and the sample mean of a sample of size 75 is $\bar{x} = 1.1945$. Calculating $\theta = -(x_\ell - \bar{x})/\sigma$, we obtain $\theta = 1.417$ as in Example 3.1 and thus $\delta = 0.22055$. The maximum likelihood estimator is given by

$$\begin{aligned}\mu_{ML} &= \bar{x} - \delta\sigma = 1.1945 - 0.22055 \times 0.006 \\ &= 1.193177\end{aligned}$$

The variance of μ_{ML} is given by (3.14) replacing $\hat{u}_r$ by $-u_\ell$. Hence we obtain the same variance as in Example 3.1.

σ^2 unknown. If σ^2 is unknown we use the following transformation which is due to Cohen (1959). Let

$$\omega = \frac{s^2}{(x_r - \bar{x})^2} \tag{3.15}$$

where $\bar{x}$ and s^2 are the sample mean and sample variance; then the equations (3.7) and (3.8) can be reduced to an equation for u_r:

$$\omega = \frac{1 - W(u_r)[W(u_r) + u_r]}{[W(u_r) + u_r]^2} \tag{3.16}$$

Let

$$Q(\omega) = \frac{W(u_r)}{W(u_r) + u_r}$$

where u_r is the unique solution to equation (3.16); then the ML-estimators can be written as Cohen's (1959) estimators which have the form

$$\sigma^2_{ML} = s^2 + Q(\omega)(x_r - \bar{x})^2 \tag{3.17}$$

$$\mu_{ML} = \bar{x} + Q(\omega)(x_r - \bar{x}) \tag{3.18}$$

Equation (3.16) might be solved by the Newton-Raphson iteration method. To avoid iteration procedures the following rational approximation was derived which can be used when the truncation is less than 50% of the population.

Rational approximation. Several rational functions were tested. The following procedure was found to be most appropriate for practical purposes. Truncation higher than 50% corresponds to $\omega > 0.57081$.

1. If $\omega < 0.06246$ (corresponds to $u_r > 4$), then set $Q(\omega) = 0$. The maximal |error| is less than 10^{-5}.
2. If $0.06246 \leqslant \omega \leqslant 0.57081$ then set

$$Q(\omega) = P_4(\omega)/P_3(\omega) + \epsilon(\omega)$$

$$P_4(\omega) = -0.00374615 + 0.17462558\omega - 2.87168509\omega^2 + 17.48932655\omega^3 - 11.91716546\omega^4$$

$$P_3(\omega) = 1 + 5.74050101\omega - 13.53427037\omega^2 + 6.88665552\omega^3$$

The maximal |error|, $|\epsilon(\omega)| \leqslant 5 \times 10^{-6}$.

No approximation will be given for $\omega > 0.57081$, since the variances of the estimators become extremely high (see Section 3.3). Tables and a nomograph for determining $Q(\omega)$ are given in the Appendix A.

A similar nomograph and tables were prepared by Cohen (1959). Chiu and Leung (1981) presented an alternate graphical method.

Asymptotic variance. The variance and covariance of the estimators μ_{ML} and σ_{ML} are obtained by inverting the Fisher information matrix

$$nJ = \begin{bmatrix} -E[\partial^2 \log L/\partial^2\mu] & -E[\partial^2 \log L/\partial\mu\,\partial\sigma] \\ -E[\partial^2 \log L/\partial\mu\,\partial\sigma] & -E[\partial^2 \log L/\partial^2\sigma] \end{bmatrix} = \frac{n}{\sigma^2}\begin{bmatrix} J_{11} & J_{12} \\ J_{21} & J_{22} \end{bmatrix}$$

where the elements J_{ij} are given by

$$J_{11} = 1 - W(u_r)[W(u_r) + u_r] \tag{3.19}$$

$$J_{12} = J_{21} = W(u_r)\{1 + u_r[W(u_r) + u_r]\} \tag{3.20}$$

$$J_{22} = 2 + u_r J_{12} \tag{3.21}$$

The asymptotic covariance matrix (ASCV) is thus

$$\mathrm{ASCV}(\mu_{ML}, \sigma_{ML}) = \frac{\sigma^2}{n} \frac{1}{J_{11}J_{22} - J_{12}^2} \begin{bmatrix} J_{22} & -J_{12} \\ -J_{12} & J_{11} \end{bmatrix} \tag{3.22}$$

It was tabulated by Cohen (1961) for some selected truncations. For convenience we have prepared Table A.3 in Appendix A.

Example 3.3

We reconsider Example 3.1; we assume, however, that the variance σ^2 is unknown and has to be estimated.

Let

$$x_r = 1.203$$

$$\bar{x} = 1.1945$$

$$s^2 = 3.6 \times 10^{-5}$$

We calculate $\omega = 0.4983$ and read from Figure A2 of Appendix A the value $Q(\omega) \simeq 0.6$. Table 1 in the appendix gives a more accurate value of $Q(\omega) \simeq 0.58864$ (interpolated). Thus the maximum likelihood estimators are

$$\sigma^2_{ML} = 0.000036 + 0.58864 \times 0.0000723 = 0.0000786$$

$$\mu_{ML} = 1.1945 + 0.58864 \times 0.0085 = 1.1995$$

The rational approximation (2) results in $Q(\omega) = 0.59113$, but the estimates do not differ significantly from the ones obtained above.

This example shows the effect an estimation of σ has on the estimate μ_{ML}. In this example it turns out that in the case where $\hat{\sigma}$ has to be estimated, the maximum likelihood estimator of μ is significantly higher than for the case where σ is known. Standard errors can be calculated by evaluating equation (3.22). But we mention that those standard errors have to be interpreted very carefully.

The standard approach of considering a modified t-statistic as approximately normally distributed does not work since $\hat{\mu}$ and $\hat{\sigma}$ are highly correlated for moderate truncation. The distribution of t is therefore skewed for finite sample sizes.

The derivation of confidence intervals for μ is aggravated by the fact that ASCV(μ_{ML}, σ_{ML}) has to be estimated itself.

Example 3.4

Cohen (1959) gives an example of left truncation. In order to insure meeting a lower specification limit of 0.1215 inches on the thickness of a certain insulation washer, all production of this component is sorted through go, no-go gauges, and all of thickness less than this value are discarded. For a random sample of 100 washers selected from the screened production, it is found that $\bar{x} = 0.124624$ and $s^2 = 2.1106 \times 10^{-6}$. Since n = 100 and $x_\ell = 0.1215$, then $(\bar{x} - x_\ell)^2 = 0.21627$. By linear interpolation in Table A1, we obtain $Q(\omega) = 0.030137$. The value obtained by the rational approximation is 0.030118 and this value might also have been read from Figure A2 of Appendix A to two decimal places as 0.03. Under the assumption that X is normally distributed, the estimates of μ and σ^2 are

$$\begin{aligned} \sigma^2_{ML} &= 2.1106 \times 10^{-6} + 0.030118(0.003124)^2 \\ &= 2.405 \times 10^{-6} \\ \mu_{ML} &= 0.214624 - 0.030118(0.003124) \\ &= 0.12453 \end{aligned}$$

The Doubly Truncated Normal Distribution

Let X be a normally distributed random variable truncated at the right point x_r and left point x_ℓ. The pdf is

$$f(x;\mu,\sigma,x_\ell,x_r) = \begin{cases} \dfrac{\phi[(x-\mu)/\sigma]}{\sigma\{\Phi[(x_r-\mu)/\sigma] - \Phi[(x_\ell-\mu)/\sigma]\}} & x_\ell \leq x \leq x_r \\ 0 & \text{otherwise} \end{cases} \tag{3.23}$$

The truncation points are assumed to be known.

Suppose a sample $(x_1, \ldots, x_n)$ of size n is given from a population with pdf $f(x; \mu, \sigma, x_\ell, x_r)$. With $u_r = (x_r - \mu)/\sigma$ and $u_\ell = (x_\ell - \mu)/\sigma$ the likelihood function is

$$L(\mu,\sigma) = \{\sigma\sqrt{2\pi}\,[\Phi(u_r) - \Phi(u_\ell)]\}^{-n} \exp\left[\sum_{i=1}^{n} -(x_i - \mu)^2/2\sigma^2\right] \tag{3.24}$$

The natural logarithm of the likelihood function is given up to a constant by

$$\log L(\mu,\sigma) = -n\{\log[\Phi(u_r) - \Phi(u_\ell)]\} - n \log \sigma - 1/(2\sigma^2) \sum_{i=1}^{n} (x_i - \mu)^2 \tag{3.25}$$

If both μ and σ are unknown the estimates μ_{ML} and σ_{ML} are solutions to the likelihood equations

$$\frac{\partial \log L(\mu,\sigma)}{\mu} = \frac{n}{\sigma}\left[\frac{\phi(u_r) - \phi(u_\ell)}{\Phi(u_r) - \Phi(u_\ell)}\right] + \frac{1}{\sigma 2} \sum_{i=1}^{n} (x_i - \mu) = 0 \tag{3.26}$$

$$\frac{\partial \log L(\mu,\sigma)}{\partial \sigma} = \frac{n}{\sigma}\, \frac{\phi(u_r)u_r - \phi(u_\ell)u_\ell}{\Phi(u_r) - \Phi(u_\ell)} - \frac{n}{\sigma} + \frac{1}{\sigma^3} \sum_{i=1}^{n} (x_i - \mu)^2 = 0 \tag{3.27}$$

Using the notations Z_r and Z_ℓ defined by (2.5) and (2.6), these two equations can be written as (Cohen, 1957)

$$\frac{Z_r - Z_\ell + u_r}{u_r - u_\ell} = \frac{x_r - \bar{x}}{x_r - x_\ell} \tag{3.28}$$

$$\frac{1 - Z_r u_r + Z_\ell u_\ell - (Z_r - Z_\ell)^2}{(u_r - u_\ell)^2} = \frac{s^2}{(x_r - x_\ell)^2} \tag{3.29}$$

where $\bar{x}$ and s^2 denote the sample mean and variance respectively. Note that (3.28) and (3.29) are equations for u_r and u_ℓ. The Newton-Raphson method might be used to find solutions to the latter equations. The likelihood estimators are then

$$\mu_{ML} = x_r - \frac{u_r}{u_r - u_\ell}(x_r - x_\ell) \tag{3.30}$$

$$\sigma_{ML} = \frac{1}{u_r - u_\ell}(x_r - x_\ell) \tag{3.31}$$

σ^2 known. If σ^2 is known we define $\delta = (\mu_{ML} - \bar{x})/\sigma$ and obtain from (3.28) the equation

$$\delta = Z_r\left(\frac{x_\ell - \bar{x}}{\sigma} - \delta, \frac{x_r - \bar{x}}{\sigma} - \delta\right) - \left(Z_\ell \frac{x_\ell - \bar{x}}{\sigma} - \delta, \frac{x_r - \bar{x}}{\sigma} - \delta\right) \tag{3.32}$$

which is an equation for δ. And the maximum likelihood estimator is thus

$$\mu_{ML} = \bar{x} + \delta\sigma \tag{3.33}$$

The Newton-Raphson method can be used to solve equation (3.32). For light truncation or nearly symmetrical truncation the following Taylor approximation can be applied. Let $\theta_r = (x_r - \bar{x})/\sigma$ and $\theta_\ell = (x_\ell - \bar{x})/\sigma$; then we obtain

$$\delta = \frac{Z_r - Z_\ell}{1 + Z'_r - Z'_\ell} \tag{3.34}$$

which results from a Taylor expansion and truncation after the linear term. The derivatives Z'_r and Z'_ℓ are

$$Z'_r = -\theta_r[Z_r + Z_r(Z_r - Z_\ell)] \tag{3.35}$$

$$Z'_\ell = \theta_\ell[Z_\ell - Z_\ell(Z_r - Z_\ell)] \tag{3.36}$$

The expressions Z_ℓ and Z_r are evaluated at θ_r and θ_ℓ given above, i.e.,

$$Z_r = Z_r\left(\frac{x_\ell - \bar{x}}{\sigma}, \frac{x_r - \bar{x}}{\sigma}\right) \tag{3.37}$$

$$Z_\ell = Z_\ell\left(\frac{x_\ell - \bar{x}}{\sigma}, \frac{x_r - \bar{x}}{\sigma}\right) \tag{3.38}$$

<u>σ^2 unknown.</u> We first describe the Newton-Raphson method applied to the likelihood equations (3.26) and (3.27). Let $[\mu^{(k)}, \sigma^{(k)}]$ be the kth iteration with the Newton-Raphson method; then

$$\begin{bmatrix} \mu^{(k+1)} \\ \sigma^{(k+1)} \end{bmatrix} = \begin{bmatrix} \mu^{(k)} \\ \sigma^{(k)} \end{bmatrix} - J^{(k)^{-1}} \begin{bmatrix} \partial \log L/\partial \mu \\ \partial \log L/\partial \sigma \end{bmatrix}^{(k)}$$

where $J^{(k)}$ is the matrix of the second derivatives evaluated at $[\mu^{(k)}, \sigma^{(k)}]$ and the first derivatives at the right hand side are evaluated at $[\mu^{(k)}, \sigma^{(k)}]$ as well. One of the advantages of this method is that the inverse of the matrix $J^{(k)}$, which is computed in each iteration step, can be used as an estimate of the covariance matrix.

Let J_{ij} be the ijth element of the matrix $J^{(k)}$; then

$$J^{(k)^{-1}} = \frac{1}{J_{11}J_{22} - J_{12}^2} \begin{bmatrix} J_{22} & -J_{12} \\ -J_{12} & J_{11} \end{bmatrix}$$

and $[\mu = \mu^{(k)}, \sigma = \sigma^{(k)}]$

$$J_{11} = \frac{n}{\sigma^2}[Z_r u_r + (Z_r - Z_\ell)^2 - Z_\ell u_\ell - 1]$$

$$J_{22} = \frac{n}{\sigma^2}\left\{Z_r u_r^3 + (Z_r u_r - Z_\ell u_\ell)^2 - Z_\ell u_\ell^3 + 1 \right.$$
$$\left. -2(Z_r u_r - Z_\ell u_\ell) + \frac{s^2}{\sigma^2} + \left[\frac{\bar{x} - \mu}{\sigma}\right]^2\right\}$$

$$J_{12} = \frac{n}{\sigma^2}\left[Z_r u_r^2 + (Z_r - Z_\ell)(Z_r u_r - Z_\ell u_\ell) \right.$$
$$\left. - (Z_r - Z_\ell) - Z_\ell u_\ell^2 - 2\,\frac{\bar{x} - \mu}{\sigma}\right]$$

where Z_r and Z_ℓ are given by formula (2.5) and (2.6) respectively.

Instead of solving equations (3.26) and (3.27) we might also solve (3.28) and (3.29) by the Newton-Raphson method in order to find u_r and u_ℓ. This was suggested by Cohen (1957), who also prepared a nomograph for finding u_r and u_ℓ.

A different approach is introduced here, which leads to another form of nomograph. We define

$$H = \frac{1}{u_r - u_\ell} \tag{3.39}$$

$$G = \frac{u_r}{u_r - u_\ell} \tag{3.40}$$

then

$$\mu = x_r - G(x_r - x_\ell)$$

and

$$\sigma = H(x_r - x_\ell)$$

Furthermore let

$$a = \frac{x_r - \bar{x}}{x_r - x_\ell} \tag{3.41}$$

and

$$b = \frac{s^2}{(x_r - x_\ell)^2} \tag{3.42}$$

then the likelihood equations for G and H are

$$H(Z_r - Z_\ell) + G - a = 0 \tag{3.43}$$

and

$$H^2 - Z_\ell H + 2Ga - G^2 - HG(Z_r - Z_\ell) = 0$$

Using equation (3.43), the latter equation can be simplified to

$$H^2 - Z_\ell H + aG - a^2 - b = 0 \tag{3.44}$$

which are equations for H and G. Note that Z_ℓ [G/H, (1 - G)/H] and Z_r[G/H, (1 - G)/H]. The maximum likelihood estimators of μ and σ are thus

$$\mu_{ML} = x_r - G(x_r - x_\ell) \tag{3.45}$$

$$\sigma_{ML} = H(x_r - x_\ell) \tag{3.46}$$

A Fortran IV program based on the Newton-Raphson method using equations (3.45) and (3.46) is presented in Appendix B.

We have prepared a nomograph to find H and G, given $0.2 \leq a \leq 0.8$ and $0.015 \leq b \leq 0.073$. Due to the symmetry of the normal distribution we only have to plot the nomograph for values $0.2 \leq a \leq 0.5$. If $a > 0.5$, set $a = 1 - a$ and read G - 1 on the abscissa. The nomograph is given in Appendix A, Figure 3.

The nomograph provides solutions only for $x_\ell \leq \mu \leq x_r$. But even if this condition holds, samples of small sizes might result in a and b such that G becomes negative and thus μ_{ML} lies above x_r. In this case the Newton-Raphson method usually diverges, because the iteration matrix is ill-conditioned and appropriate starting values are difficult to assign.

If $\mu_{ML} < x_\ell$ or $\mu_{ML} > x_r$, the appropriateness of the normal model should be reconsidered, since the variances of the estimators become enormous. To avoid situations where $\mu_{ML} < x_\ell$ or $\mu_{ML} > x_r$, even though the true value μ lies within the interval (x_ℓ, x_r), the sample size has to be increased.

Cohen's (1957) nomograph for u_r and u_ℓ avoids scaling problems which appear in the nomograph given in Figure 3. But, on the other hand, a nomograph for H and G better reflects the sensitivity (and variance) of the solutions for increasing b. Slight changes of b and a cause (for large b) drastic changes in $H = 1/(u_r - u_\ell)$ and $G = u_r/(u_r - u_\ell)$ and thus drastic changes in the estimates σ_{ML} and μ_{ML}. For instance, if b = 0.067 and a = 0.46, then G = 0.4 and H = 0.407. If b = 0.069 and a = 0.44, then G = 0.32 and H = 0.457. Thus the estimate $\hat{\sigma} = (x_r - x_\ell)H$ of the standard deviation increases about 12% while the estimate of $x_r - \hat{\mu}$ increases about 20%.

An alternate graphical method for estimating the parameters of a truncated normal distribution was considered by Chiu and Leung (1981).

Example 3.5

We consider Example 3.3; we assume, however, that all items of diameter less than 1.75 inches and above 1.207 inches are discarded. Thus

$$x_\ell = 1.175$$

$$x_r = 1.207$$

$$\bar{x} = 1.1945$$

$$s^2 = 3.6 \times 10^{-5}$$

We calculate $d = x_r - x_\ell = 0.032$

$$a = (x_r - \bar{x})/d = 0.3906$$

$$b = s^2/d^2 = 0.0352$$

From Figure A3 of Appendix A we obtain H = 0.205 and G = 0.375 and therefore (3.45) and (3.46) give

$$\mu_{ML} = 1.195$$

$$\sigma_{ML} = 0.00656$$

Asymptotic variance. The asymptotic covariance matrix of the MLE is given by the inverse of the Fisher information matrix J. Let

$$J_{11} = 1 + u_\ell Z_\ell - u_r Z_r - (Z_\ell - Z_r)^2 \tag{3.47}$$

$$J_{12} = J_{21} = (Z_r - Z_\ell)(1 + u_r Z_r - u_\ell Z_\ell) + u_r^2 Z_r - u_\ell^2 Z_\ell \tag{3.48}$$

$$J_{22} = 2 - (u_r Z_r - u_\ell Z_\ell)^2 + u_\ell Z_\ell - u_r Z_r + u_\ell^3 Z_\ell - u_r^3 Z_r \tag{3.49}$$

then

$$\mathrm{ASCV}(\mu_{ML}, \sigma_{ML}) = \frac{\sigma^2}{n} \frac{1}{J_{11}J_{22} - J_{12}^2} \begin{bmatrix} J_{22} & -J_{12} \\ -J_{12} & J_{11} \end{bmatrix} \qquad (3.50)$$

The EM algorithm. The expectation maximization (EM) algorithm was proposed by Dempster, Laird, and Rubin (1977) to find estimates for truncated samples. Let n be the size of the sample, excluding the unknown number m of truncated points. Suppose m_ℓ observations are truncated from the left and m_r observations are truncated from the right. Given m_ℓ, we assume that the truncated sample values $\tilde{x}_\ell = (x_{\ell 1}, \ldots, x_{\ell m_\ell})$ are a random sample of size m_ℓ from the density of $\sigma^{-1}\phi[(x - \mu)/\sigma]/\Phi[(x_\ell - \mu)/\sigma]$. The same holds for m_r; we suppose $\tilde{x}_r = (x_{r1}, \ldots, x_{rm_r})$ is a random sample of size m_r from the density $\sigma^{-1}\phi[(x - \mu)/\sigma]/\{1 - \Phi[(x_r - \mu)/\sigma]\}$.

In our specific case, the EM algorithm is an iterative procedure which consists of two steps. Suppose that $(\hat{\mu}^i, \hat{\sigma}^i)$ denotes the current estimate of (μ, σ) after i cycles of the algorithm. Then the next cycle can be described in two steps.

E-step: Estimate the complete data sufficient statistic by finding

$$E[x \mid x < x_\ell, \hat{\mu}^i, \hat{\sigma}^i], \quad E[m_\ell \mid x_\ell, x_r, \hat{\mu}^i, \hat{\sigma}^i]$$
$$E[x^2 \mid x < x_\ell, \hat{\mu}^i, \hat{\sigma}^i] \text{ and}$$
$$E[x \mid x > x_r, \hat{\mu}^i, \hat{\sigma}^i], \quad E[m_r \mid x_\ell, x_r, \hat{\mu}^i, \hat{\sigma}^i]$$
$$E[x^2 \mid x < x_r, \hat{\mu}^i, \hat{\sigma}^i]$$

M-step: Determine $\hat{\mu}^{i+1}$ and $\hat{\sigma}^{i+1}$ by ordinary maximum likelihood regarding the components $(x_1, \ldots, x_n, x_{\ell 1}, \ldots, x_{\ell m_\ell}, x_{r1}, x_{rm_r})$ as a random sample of size $n + m_\ell + m_r$ from $\sigma^{-1}\phi[(x - \mu)/\sigma]$, where $x_{\ell j}$ $(j = 1, \ldots, m_\ell)$ and x_{rj} $(j = 1, \ldots, m_r)$ are replaced by their expectations.

The M-step in our case only requires the calculation of

$$\hat{\mu}^i = \sum_{j=1}^{n+m} x_j/(n+m) \qquad m = m_\ell + m_r$$

$$\hat{\sigma}^i = \left[\sum_{j=1}^{n+m} (x_j - \bar{x})^2/(n+m) \right]^{1/2} \qquad m = m_\ell + m_r$$

where x_j ($j = 1, \ldots, n$) are the observed values and x_j ($j = n + 1, \ldots, n + m$) are the expected values given the estimates (μ^{i-1}, σ^{i-1}) of cycle $i - 1$. But note that in the last equation x_j^2 is replaced by $E[x^2 \mid x > x_\ell, \hat{\mu}^{i-1}, \hat{\sigma}^{i-1}]$ when truncated on the left. Straightforward evaluations yield

$$\hat{m}_r = n\phi(\hat{u}_\ell)/[\Phi(\hat{u}_r) - \Phi(\hat{u}_\ell)]$$

$$\hat{m}_\ell = n[1 - \phi(\hat{u}_r)]/\phi(\hat{u}_r) - \phi(\hat{u}_\ell)]$$

$$\hat{x}_{rj} = \hat{\mu} + \sigma W(-\hat{u}_r) \qquad j = 1, \ldots, \hat{m}_r$$

$$\hat{x}_{\ell j} = \hat{\mu} - \sigma W(\hat{u}\ \ ) \qquad j = 1, \ldots, \hat{m}_\ell$$

$$\hat{x}_{rj}^2 = \hat{\sigma}^2[1 + \hat{u}_r W(-\hat{u}_r)] \qquad j = 1, \ldots, \hat{m}_r$$

$$\hat{x}_{\ell j}^2 = \hat{\sigma}^2[1 - \hat{u}_\ell W(\hat{u}_\ell)] \qquad j = 1, \ldots, \hat{m}_\ell$$

where $\hat{u}_\ell$, $\hat{u}_r$, $\hat{\mu}$, and $\hat{\sigma}$ are the estimates of the present cycle.

The EM algorithm has the advantage that it converges, but the convergence can become very slow. The number of iterations needed to meet the convergence criterion is usually higher than for the Newton-Raphson algorithm and the number of iterations for the EM algorithm is over 60 when the Newton-Raphson algorithm does not converge. Simulation results will therefore not be reported.

Restricted maximum likelihood estimator. If it is known that the parameter μ lies within the interval (x_ℓ, x_r), then restricted maximum likelihood estimators should be used. Although the restricted MLEs are more biased, their mean square error is significantly lower compared with the MLE.

The restricted maximum likelihood estimators in the singly right truncated case are

$$\hat{\mu} = \begin{cases} \mu_{ML} & \text{if } \mu_{ML} < x_r \\ x_r & \text{if } \mu_{ML} \geqslant x_r \end{cases}$$

and

$$\hat{\sigma}^2 = \begin{cases} \sigma^2_{ML} & \mu_{ML} < x_r \\ \dfrac{1}{n}\displaystyle\sum_{i=1}^{n} (x_i - x_r)^2 & \mu_{ML} \geqslant x_r \end{cases}$$

These results are derived from (3.2) and (3.3).

In the doubly truncated case the restricted MLE $\hat{\mu}$ takes on the value x_r or x_ℓ, respectively, if μ_{ML} lies outside the interval (x_ℓ, x_r). From (3.27) we then obtain an equation for σ

$$-\sigma^2 + \sigma d \frac{\Phi(d/\sigma)}{0.5 + \Phi(d/\sigma)} + \frac{1}{n}\sum_{i=1}^{n} (x_i - \hat{\mu})^2 = 0$$

3.2 MOMENT ESTIMATORS

The method of maximum likelihood and the method of moments yield the same estimators. The method of moments based on the first two moments does not thus have any advantages over the maximum likelihood method. But since the likelihood equations are usually solvable only by iterative methods, Cohen (1950b) and Shah and Jaiswal (1966) have suggested using three and four moments, respectively, to obtain explicit results for the estimators. Cohen considers the singly truncated normal distribution while Shah and Jaiswal derive estimators for the doubly truncated case. Those estimators have of course higher variances than the ML-estimators. In 3.1 we have shown that for the singly truncated normal distribution we are able to give explicit results by using rational function approximation. The moment estimators have the disadvantage that they can result in negative variance estimates.

The following results were obtained by Cohen (1950b) and Shal and Jaiswal (1966).

Singly Truncated Normal Distribution

Consider a singly left truncated normal distribution with known truncation point x_ℓ. The moments around x_ℓ were given in Chapter 2. Let $h = -u\sigma$, where $u_\ell = (x_\ell - \mu)/\sigma$; then the first three moments are

$$m_1 = \sigma W(h/\sigma) + h$$

$$m_2 = \sigma^2 + hm_1$$

$$m_3 = hm_2 + 2\sigma^2 m_1$$

The last two equations can be solved for h and σ^2

$$h = \frac{2m_1 m_2 - m_3}{2m_1^2 - m_2} \tag{3.51}$$

$$\sigma^2 = \frac{m_1 m_3 - m_2^2}{2m_1^2 - m_2} \tag{3.52}$$

We now substitute m_k by the kth sample moment around x_ℓ

$$\hat{m}_k = \frac{1}{n}\sum_{i=1}^{n}(x_i - x_\ell)^k$$

and obtain estimators σ_{MM} and μ_{MM} where

$$\mu_{MM} = x_\ell + \hat{h} \tag{3.53}$$

<u>σ^2 known.</u> Suppose σ^2 is known; then we obtain with

$$m_2 = \sigma^2(1 + hm_1)$$

the formula for μ_{MM}

$$\mu_{MM} = x_\ell + \frac{\sigma^2 - m_2}{m_1} \tag{3.54}$$

The asymptotic variances of μ_{MM} and σ_{MM} are obtained by the delta method (Cohen, 1950b).

$$\text{var}(\mu_{MM}) = \sum_{i=1}^{3}\sum_{j=1}^{3} A_i A_j \,\text{cov}(\hat{m}_i, \hat{m}_j) \tag{3.55}$$

$$\text{var}(\sigma_{MM}) = \sum_{i=1}^{3}\sum_{j=1}^{3} B_i B_j \,\text{cov}(\hat{m}_i, \hat{m}_j) \tag{3.56}$$

where

$$A_j = \left.\frac{\partial \mu_{MM}}{\partial \hat{m}_j}\right|_{\hat{m}_j = m_j} \qquad j = 1, 2, 3$$

$$B_j = \left.\frac{\partial \sigma_{MM}}{\partial \hat{m}_j}\right|_{\hat{m}_j = m_j} \qquad j = 1, 2, 3$$

We now proceed to evaluate A_j and B_j. Let

$$\omega_j = m_j/\sigma^j$$

hence

$$\omega_1 = W(-u_\ell) - u_\ell$$

$$\omega_2 = 1 - u_\ell \omega_1$$

$$\omega_3 = 2\omega_1 - u_\ell \omega_2$$

$$\omega_j = (j-1)\omega_{j-2} - u_\ell \omega_{j-1} \qquad j = 4, 5, 6$$

Straightforward calculations yield

$$A_1 = \frac{1}{C}(2\omega_2 + 4u_{\ell}\omega_1)$$

$$A_2 = \frac{1}{\sigma C}(2\omega_1 - u_{\ell})$$

$$A_3 = -\frac{1}{\sigma^2 C}$$

$$B_1 = \frac{1}{2C}(\omega_3 - 4\omega_1)$$

$$B_2 = \frac{1}{2\sigma C}(1 - 2\omega_2)$$

$$B_3 = \frac{1}{2\sigma^2 C}\omega_1$$

where

$$C = 2\omega_1^2 - \omega_2$$

Furthermore, we have

$$\text{var}(\hat{m}_i) = \frac{1}{n}(m_{2i} - m_i^2) \tag{3.57}$$

$$\text{cov}(\hat{m}_i, \hat{m}_j) = \frac{1}{n}(m_{i+j} - m_i m_j) \tag{3.58}$$

Let $a_j = A_j\sigma^{j-1}$ and $b_j = B_j\sigma^{j-1}$; then

$$\frac{n}{\sigma^2}\text{var}(\mu_{MM}) = \sum_{i=1}^{3}\sum_{j=1}^{3}(\omega_{i+j} - \omega_i\omega_j)a_i a_j$$

$$\frac{n}{\sigma^2}\text{var}(\sigma_{MM}) = \sum_{i=1}^{3}\sum_{j=1}^{3}(\omega_{i+j} - \omega_i\omega_j)b_i b_j$$

If σ^2 is known we obtain

$$A_1 = \frac{1}{\omega_1}(\omega_2 - 1)$$

$$A_2 = \frac{1}{3\omega_1}$$

$$A_2 = A_3 = 0$$

Doubly Truncated Normal Distribution

Consider a doubly truncated normal distribution with truncation points x_ℓ and x_r. The moments around x_ℓ are given in Chapter 2. Let $d = x_r - x_\ell$, then the first four moments can be written as

$$m_1 = \sigma(Z_\ell - Z_r) - \sigma u_\ell$$

$$m_2 = -\sigma d Z_r + \sigma^2 - \sigma u_\ell m_1$$

$$m_3 = -\sigma d^2 Z_r + 2\sigma^2 m_1 - \sigma u_\ell m_2$$

$$m_4 = -\sigma d^3 Z_r + 3\sigma^2 m_2 - \sigma u_\ell m_3$$

Let $h = -\sigma u_\ell$, then the four equations can be solved for h and σ^2

$$h = -\frac{3m_2m_3 - 2m_1m_4 + d(m_4 - 3m_2^2) + d^2(2m_1m_2 - m_3)}{(2m_1m_3 - 3m_2^2) + d(3m_1m_2 - m_3) + d^2(m_2 - 2m_1^2)} \tag{3.59}$$

$$\sigma^2 = \frac{(m_3^2 - m_2m_4) + d(m_1m_4 - m_2m_3) + d^2(m_2^2 - m_1m_3)}{(2m_1m_3 - 3m_2^2) + d(3m_1m_2 - m_3) + d^2(m_2 - 2m_1^2)} \tag{3.60}$$

We substitute the moments m_k by the sample moments around x_ℓ

$$\hat{m}_k = \frac{1}{n}\sum_{i=1}^{n}(x_i - x_\ell)^k$$

and thereby obtain explicit results for μ_{MM} and σ^2_{MM}.

σ^2 known. If σ^2 is known we obtain

$$h = \frac{m_3 + dm_2 - 2\sigma^2 m_1 - d\sigma^2}{dm_1 + m_2} \tag{3.61}$$

and

$$\mu_{MM} = x_\ell + h \tag{3.62}$$

The asymptotic variances of μ_{MM} and σ_{MM} are obtained by the delta method (Shah and Jaiswal, 1966).

$$\text{var}(\mu_{MM}) = \sum_{i=1}^{4}\sum_{j=1}^{4} A_i A_j \,\text{cov}(\hat{m}_i, \hat{m}_j)$$

$$\text{var}(\sigma_{MM}) = \sum_{i=1}^{4}\sum_{j=1}^{4} A_i A_j \,\text{cov}(\hat{m}_i, \hat{m}_j)$$

where

$$A_i = \left.\frac{\partial \hat{h}}{\partial \hat{m}_i}\right|_{\hat{m}_i = m_i} \qquad i = 1, 2, 3, 4$$

$$B_i = \left.\frac{\partial \hat{\sigma}}{\partial \hat{m}_i}\right|_{\hat{m}_i = m_i} \qquad i = 1, 2, 3, 4$$

$$\text{var}(\hat{m}_i) = (m_{2i} - m_i^2)/n$$

$$\text{cov}(\hat{m}_i, \hat{m}_j) = (m_{i+j} - m_i m_j)/n$$

Let $m_j = \sigma^j \omega_j$; then

$$\omega_1 = Z_1 - Z_r - u_\ell$$

$$\omega_2 = -Z_r(u_r - u_\ell) + 1 - u_\ell \omega_1$$

$$\omega_j = -Z_r(u_r - u_\ell)^{j-1} + (j-1)\omega_{j-2} - u_\ell\,\omega_{j-1} \qquad j = 3, 4, \ldots, 8$$

We obtain

$$A_1 = -[-2(\omega_4 + u_\ell\omega_3) - 3u_\ell\omega_2(u_r - u_\ell) + 2(u_r - u_\ell)^2(2u_\ell\omega_1 + \omega_2)]/C$$

$$A_2 = -[3(\omega_3 + 2u_\ell\,\omega_2) - 3(u_r - u_\ell)(2\omega_2 + u_\ell\,\omega_1) + (u_r - u_\ell)^2 \times (2\omega_1 - u_\ell)]/\sigma C$$

$$A_3 = -[(3\omega_2 - 2u_\ell\,\omega_1) + u_\ell(u_r - u_\ell) - (u_r - u_\ell)^2]/\sigma^2 C$$

$$A_4 = -[-2\omega_1 + (u_r - u_\ell)]/\sigma^3 C$$

$$B_1 = [-2\omega_3 + (u_r - u_\ell)(\omega_4 - 3\omega_2) + (u_r - u_\ell)^2(4\omega_1 - \omega_3)]/2C$$

$$B_2 = [(6\omega_2 - \omega_4) - (u_r - u_\ell)(\omega_3 + 3\omega_1) + (u_r - u_\ell)^2(2\omega_2 - 1)]/2\sigma C$$

$$B_3 = [2(\omega_3 - \omega_1) + (u_r - u_\ell)(1 - \omega_2) - \omega_1(u_r - u_\ell)^2]/2\sigma^2 C$$

$$B_4 = [-\omega_2 + \omega_1(u_r - u_\ell)]/2\sigma^3 C$$

where

$$C = (2\omega_1\omega_3 - 3\omega_2^2) + (3\omega_1\omega_2 - \omega_3)(u_r - u_\ell) + (\omega_2 - 2\omega_1^2)(u_r - u_\ell)^2$$

Let $a_j = A_j\sigma^{j-1}$, $b_j = B_j\sigma^{j-1}$; then we finally obtain

$$\frac{n}{\sigma^2}\,\mathrm{var}(\mu_{MM}) = \sum_{i=1}^{4}\sum_{j=1}^{4}(\omega_{i+j} - \omega_i\omega_j)a_i a_j \tag{3.63}$$

$$\frac{n}{\sigma^2}\,\mathrm{var}(\sigma_{MM}) = \sum_{i=1}^{4}\sum_{j=1}^{4}(\omega_{i+j} - \omega_i\omega_j)b_i b_j \tag{3.64}$$

σ^2 known. If σ^2 is known, then

$$\omega_1 = Z_1 - Z_r - u_\ell$$

$$\omega_2 = -Z_r(u_r - u_\ell) + 1 - u_\ell \omega_1$$

$$\omega_i = -Z_r(u_r - u_\ell)^{j-1} + (j - 1)\omega_{j-2} - u_r\omega_{j-1} \qquad j = 3, 4, 5, 6$$

and we obtain

$$A_1 = \frac{u_\ell(u_r - u_\ell) - 2}{C}$$

$$A_2 = \frac{u_r}{\sigma C}$$

$$A_3 = \frac{1}{\sigma^2 C}$$

$$C = (u_r - u_\ell)\omega_1 + \omega_2$$

3.3 BIAS CORRECTIONS

The ML-estimators as well as the MM-estimators are biased. The bias varies with the magnitude of truncation. Fergeson et al. (1978) have investigated the effect that jackknifing the ML-estimates has on bias and variance for sample size $n = 50$. This investigation shows that moderate levels of truncation are needed before worthwhile reductions occur.

The jackknife estimators are derived by the following procedure. Let $x_1, \ldots, x_n$ be a sample of an independent random variable from a truncated normal distribution. We consider the ML-estimators μ_{ML} and σ_{ML} based on the sample of size n. Let μ^i_{ML} and σ^i_{ML} be the corresponding estimators based on the sample of size $n - 1$, where the ith value has been deleted. Define

$$\tilde{\mu}_i = n\mu_{ML} - (n - 1)\mu^i_{ML} \qquad i = 1, \ldots, n$$

$$\tilde{\sigma}_i = n\sigma_{ML} - (n - 1)\sigma^i_{ML} \qquad i = 1, \ldots, n$$

The estimators

$$\tilde{\mu} = \frac{1}{n}\sum_{i=1}^{n} \tilde{\mu}_i$$

$$\tilde{\sigma} = \frac{1}{n}\sum_{i=1}^{n} \tilde{\sigma}_i$$

have the property that enables them to eliminate the order 1/n terms from a bias of the form

$$E[\mu_{ML}] = \mu + a_1/n + O(1/n^2)$$

$$E[\sigma_{ML}] = \sigma + b_1/n + O(1/n^2)$$

3.4 ESTIMATORS OF THE TRUNCATION POINTS

Consider a random variable X with pdf $f(x; \mu, \sigma, x_\ell, x_r)$ given by (3.23). The truncation points x_ℓ and x_r are now assumed to be unknown. Let $x_{(1)}, \ldots, x_{(n)}$ be an ordered sample of size n drawn from a population with pdf $f(x; \mu, \sigma, x_\ell, x_r)$. The maximum likelihood estimates of x_ℓ and x_r are the order statistics $x_{(1)}$ and $x_{(n)}$, respectively. This is easily seen by considering the likelihood function (3.24), which is proportional to $1/\{\Phi[(x_r - \mu)/\sigma] - \Phi[(x_\ell - \mu)/\sigma]\}$. Hence the likelihood function takes its maximum when the denominator $\Phi[(x_r - \mu)/\sigma] - \Phi[(x_\ell - \mu)/\sigma]$ takes its minimum. The latter is obviously the case when $\hat{x}_r = x_{(n)}$ and $\hat{x}_\ell = x_{(1)}$, respectively.

Charernkavanich (1978) considers modified maximum likelihood estimators. For singly left truncation the estimation equations are

$$\partial \log L/\partial \mu = 0 \qquad \partial \log L/\partial \sigma = 0$$

and

$$\hat{x}_\ell = E[x_{(1)} | x_\ell]$$

The last equation can be written as

$$\hat{x}_{\ell} = \mu + \sigma E[u_{(1)} | u_{\ell}]$$

where $u_{\ell} = (x_{\ell} - \mu)/\sigma$ and $u_{(1)} = [x_{(1)} - \mu]/\sigma$. This procedure might reduce the bias in estimating x_{ℓ}. But unfortunately a solution becomes very tedious. Not only do we have to solve three equations simultaneously but we also have to calculate the expected value of the first order statistic for a standard normal distribution which is truncated at $u_{\ell} = (\hat{x}_{\ell} - \mu)/\sigma$. Charernkavanich (1978) prepared tables for $E[u_{(1)} | u_{\ell}]$ for values of u_{ℓ} ranging from -4 to 0 and for $2 \leq n \leq 420$ to facilitate the practical application of these estimators.

The ML estimators μ_{ML} and σ_{ML} will have a higher variance if x_{ℓ} or x_r have to be estimated than when x_{ℓ} and x_r are known. In order to reveal the increase of variance we performed a simulation study with 1000 samples. The results are shown in Table 3.1.

We notice from Table 3.1 that for small sample sizes the maximum likelihood estimators μ_{ML} and σ_{ML} become heavily biased even for light truncation when the truncation points have to be estimated. Subsequently, the mean square error increases significantly. But the variance of the estimators μ_{ML} and σ_{ML} increases drastically as well. For instance, the variance of μ_{ML} for $u_{\ell} = -\infty$, $u_r = 3$ is about 2.5 times the variance of the MLE when x_{ℓ} and x_r are known. The knowledge of x_{ℓ} and x_r is particularly crucial for light truncation due to the heavy bias of $\hat{x}_{\ell}$ and $\hat{x}_r$ in those cases. If the truncation is severe, the estimators $\hat{x}_{\ell}$ and $\hat{x}_r$ become naturally less biased. We therefore conclude that it is better to ignore light truncation if the truncation points are unknown rather than to estimate the truncation points x_{ℓ} and x_r along with μ_{ML} and σ_{ML}.

3.5 ESTIMATION OF μ AND σ WHEN A KNOWN PERCENTAGE IS TRUNCATED

Consider the followind pdf

$$f(x, \mu, \sigma, p_r, x_r) = \begin{cases} \dfrac{\phi(x - \mu)/\sigma}{1 - p_r} \sigma^{-1} & x \leq x_r \\ 0 & \text{elsewhere} \end{cases} \tag{3.65}$$

TABLE 3.1 The Mean and Mean Square Error of the MLE for Unknown Truncation Points, x_ℓ and x_r (n = 20)

		μ_{ML}		σ_{ML}	
u_ℓ	u_r	Mean	MSE	Mean	MSE
$-\infty$	3.000	0.231	0.184	1.126	0.067
$-\infty$	2.000	0.221	0.155	1.109	0.067
$-\infty$	1.000	0.117	0.117	1.016	0.044
$-\infty$	0.500	-0.125	0.076	0.886	0.046
$-\infty$	0.000	-0.455	0.239	0.769	0.076
-3.000	3.000	-0.004	0.440	1.567	0.565
-3.000	2.000	0.108	0.326	1.501	0.468
-3.000	1.000	0.193	0.434	1.312	0.248
-3.000	0.500	0.076	0.352	1.155	0.135
-3.000	0.000	-0.235	0.313	0.976	0.080
-2.000	2.000	0.006	0.245	1.442	0.450
-2.000	1.000	0.055	0.282	1.210	0.187
-2.000	0.500	-0.072	0.240	1.042	0.091
-2.000	0.000	-0.374	0.291	0.829	0.080
-1.000	1.000	-0.009	0.133	0.899	0.101
-1.000	0.500	-0.153	0.112	0.704	0.133
-1.000	0.000	-0.404	0.203	0.481	0.292
-0.500	0.500	-0.001	0.045	0.492	0.280

where p_r is the proportion truncated at the right. This proportion is assumed to be known, in contrast to Sections 3.1 to 3.5. Let us first assume the truncation point is known. We will consider ML-estimators.

Suppose a sample $(x_1, x_2, \ldots, x_n)$ of size n is given from a population with pdf $f(x; \mu, \sigma, p_r, x_r)$. The joint probability density function of the sample is

$$L(\mu, \sigma) = \begin{cases} [\sigma\sqrt{2\pi}\,(1 - p_r)]^{-n} \exp\left[-\sum_{i=1}^{n} \frac{(x_i - \mu)^2}{2\sigma^2}\right] & \text{if } x_i \leqslant x_r,\ i = 1, \ldots, n \\ 0 & \text{elsewhere} \end{cases} \tag{3.66}$$

The likelihood function is dependent upon μ, σ and x_r. Since $u_r = \Phi^{-1}(1 - p_r)$ is known there is only one parameter to estimate. μ and σ are linearly dependent.

Let $\mu = x_r - \sigma u_r$; then the natural logarithm of the likelihood function is

$$\log L(\mu, \sigma) = \begin{cases} -n \log \sigma - \frac{1}{2} \sum_{i=1}^{n} \left(\frac{-(x_r - x_i)}{\sigma} + u_r\right)^2 + \text{const} & \text{if } x_i \leqslant x_r,\ i = 1, \ldots, n \\ 0 & \text{elsewhere} \end{cases} \tag{3.67}$$

Note that σ has to be unknown, otherwise we are not dealing with an estimation problem, since μ is determined by x_r, σ and u_r. The ML-estimate σ_{ML} is then the solution to

$$\frac{\partial \log L}{\partial \sigma} = -\frac{n}{\sigma} - \sum_{i=1}^{n} \left[-\frac{y_i}{\sigma} + u_r\right] \frac{y_i}{2} = 0 \tag{3.68}$$

where $y_i = x_r - x_i$. Thus σ_{ML} is solution to $\sigma^2 + \overline{y}^2 - u_r\overline{y} = 0$ and

$$\sigma_{ML} = \frac{1}{2}\left[-\overline{y}u_r = \sqrt{(\overline{y}u_r)^2 + 4\overline{y}^2}\right] \tag{3.69}$$

where

$$\overline{y} = \frac{1}{n}\sum_{i=1}^{n} y_i, \quad \overline{y}^2 = \frac{1}{n}\sum_{i=1}^{n} y_i^2, \quad u_r = \Phi^{-1}(1 - p_r) \tag{3.70}$$

We now obtain

$$\mu_{ML} = x_r - \sigma_{ML} u_r \tag{3.71}$$

If the truncated point is unknown the ML-estimator is $\hat{x}_r = x_{(n)}$, the largest order statistic.

3.6 COMPARISONS OF THE MAXIMUM LIKELIHOOD AND THE MOMENT ESTIMATORS

In this section we discuss the asymptotic efficiency of the moment estimators and the small sample performance of the MLE and of the MME. It was shown by Cohen (1950b) that in the singly truncated case the asymptotic efficiency of the moment estimator for the mean does not fall below 79%, while the asymptotic efficiency of the moment estimator for the standard deviation is greater than 74% for all truncations. Table A3 of Appendix A gives the asymptotic variances of the maximum likelihood estimators.

We also notice that the asymptotic efficiencies of the moment estimators are strongly affected by unsymmetric truncation. For instance, if $u_\ell = -0.5$ and $u_r = 5$, then the ARE of $\hat{\mu}_{MME}$ and $\hat{\sigma}^2_{MME}$ are 97% and 98%, respectively, while the efficiencies drop to 82% and 85%, respectively, if $u_r = \infty$ and $u_\ell = -0.5$. The asymptotic efficiencies in the singly truncated case can therefore serve as a lower bound for the asymptotic efficiencies for all truncations.

Asymptotic efficiencies of the moment estimators are not particularly useful for small sample sizes ($n \leq 20$) unless the truncation is light. The results of a Monte Carlo simulation study (Schneider, 1984) with sample sizes n = 10, 20, 60, 100 can be summarized as follows:

1. When the estimates of the mean lie outside the truncation interval (x_ℓ, x_r) the Newton-Raphson algorithm for calculating the maximum likelihood estimators often diverges. In these situations starting values which are accurate enough to obtain convergence are not available. But even if solutions were available, it would still be doubtful whether these solutions were meaningful because of the high variance of the estimators. The restricted maximum likelihood estimators should be used to avoid convergence problems.

TABLE 3.2 Simulation Results for the MLE and the MME for n = 20

Truncation		Mean				STD				No estimates	
		Mean		Var		Mean		Var			
Left	Right	MLE	MME	MLE	MME	MLE	MME	MLE	MME	MLE %	MME %
-3	∞	-0.007	0.032	0.053	0.049	0.967	0.900	0.029	0.117	0	0
-2	∞	-0.021	0.055	0.070	0.062	0.974	0.882	0.040	0.159	0	0
-1	∞	-0.006	0.135	0.114	0.113	0.953	0.828	0.056	0.216	0.1	2.5
0	∞	0.467	0.563	0.263	0.382	0.760	0.550	0.087	0.280	4.5	32.7
-3	3	0.003	0.035	0.054	0.049	0.973	0.914	0.033	0.031	0	0
-2	3	-0.025	0.053	0.077	0.058	0.989	0.916	0.055	0.042	0	0
-1	3	-0.095	0.123	0.324	0.106	1.000	0.891	0.108	0.064	3.3	3.7
-2	2	0.001	0.054	0.094	0.068	1.019	0.908	0.106	0.071	1.1	0.1
-1	2	-0.023	0.138	0.212	0.109	0.974	0.870	0.116	0.099	12.3	5.1
-1	1	-0.003	0.069	0.106	0.079	0.790	0.736	0.125	0.183	43.0	16.6

2. When the asymptotic variance is used for estimating the standard error of the estimates, care must be taken that the sample sizes are large enough for the degree of truncation.
3. For small sample sizes ($n \leq 20$) the formulas (3.51), (3.52), (3.59), and (3.60) are very sensitive. The numerators of (3.52) and (3.60) can have different signs than their denominators, which can result in negative variance estimates. Furthermore, in many situations the denominators become very small such that the estimates of the mean and standard deviation turn out to be extremely large. As a consequence the distributions of the MMEs are heavily skewed to the right for small sample sizes. A reasonable way to obtain more reliable estimates is therefore to use only estimates where the MME of the mean lies within the truncation limits. This procedure brings down the standard errors of the MME such that it becomes comparable with the standard errors of the MLE.
4. As a result of the restriction of the MLE and MME to the truncation interval the MLE and the MME perform similarly; but the MLE tends to have smaller bias than the MME. We notice that as an effect of the restriction the estimates can have smaller standard errors than those given by the asymptotic variance.

Some simulation results for the MLE and MME for samples of size 20 from a standard normal distribution are given in Table 3.2. These results are based on 10,000 samples taken from a standard normal distribution. The IMSL procedure GGNML was applied to generate normal variates. A restriction was used for the MME, i.e., only samples with mean inside the truncation limits were used. Thus the results for the MME are based on 10,000 samples which fulfilled the restriction and had a positive variance estimate. The results for the MLE are based on 10,000 samples for which the Newton-Raphson method converged. Note that MME% is the percentage of samples where the MME of the variance was negative or the mean was outside the truncation limits; MLE% is the percentage of samples where the Newton-Raphson method did not converge.

REFERENCES

Aitkin, M. A. and Hume, M. W. (1968). Correlation in singly truncated bivariate normal distributions, Biometrika 55, 437-441.

Charernkavanich, D. (1978). Estimation in truncated distributions when points of truncation are unknown. Dissertation, University of Georgia.

Chiu, W. K. and Leung, M. P. Y. (1981). A graphical method for estimating the parameters of a truncated normal distribution, Journal of Quality Technology 13, 42-45.

Cohen, A. C. (1949). On estimating the mean and standard deviation of truncated normal distribution, Journal of American Statistical Association 44, 518-525.

Cohen, A. C. (1950a). Estimating the mean and variance of normal populations from singly truncated and doubly truncated samples, Annals of Mathematical Statistics 21, 557-569.

Cohen, A. C. (1950b). On estimating the mean and variance of singly truncated normal frequency distributions from the first three sample moments, Annals of the Institute of Statistical Mathematics, Tokyo 1-5, 37-44.

Cohen, A. C. (1957). On the solution of estimating equations from truncated and censored samples from normal populations, Biometrika 44, 225-236.

Cohen, A. C. (1959). Simplified estimators for the normal distribution when samples are singly censored or truncated, Technometrics 1, 217-237.

Cohen, A. C. (1961). Tables for maximum-likelihood estimates: singly truncated and singly censored samples, Technometrics 3, 535-541.

Crain, B. R. (1979). Estimating the parameters of a truncated normal distribution, Applied Mathematics and Computation 5, 149-156.

Dempster, A. P., Laird, N. M., and Rubin, D. B. (1977). Maximum likelihood from incomplete data via the EM algorithm (with discussion), Journal of Royal Statistical Society, Ser. B, 39, 1-38.

Ferguson, R. A., Fryer, J. G., and McWhinney, I. A. (1978). On the estimation of a truncated normal distribution, Bulletin of the International Statistical Institute 40, 259-266.

Fisher, R. A. (1931). Properties and applications of H functions, in Mathematical Tables, Vol. 1, XXVI-XXV, British Association for the Advancement of Science.

Hald, A. (1949). Maximum likelihood estimation of the parameters of a normal distribution which is truncated at a known point, Scandinavian Actuarial Journal 32, 119-134.

Hall, R. L. (1979). Inverse moments for a class of truncated normal distributions, Sankhya B41, 66-76.

Hansen, J. N. and Zeger, S. (1980). The asymptotic variance of the estimated proportion truncated from normal population, Technometrics 22, 271-274.

Khatri, C. G. and Ratani, R. T. (1979). On estimation of the mean parameter of a truncated normal distribution with known coefficient of variation, Communications in Statistics 8, 237-244.

Pearson, K. and Lee, A. (1908). On the generalized probable error in multiple normal correlation, Biometrika 6, 59-68.

Robson, D. S. and Whitlock, J. H. (1964). Estimation of a truncation point, Biometrika 51, 33-39.

Schneider, H. (1984). Comparison of the MLE and MME for truncated samples from a normal population, Technical Report, University of Arkansas, Department of Mathematics.

Shah, S. M. and Jaiswal, M. C. (1966). Estimation of parameters of the doubly truncated normal distribution from first four sample moments, Annals of the Institute of Statistical Mathematics 18, 107-111.

CHAPTER 4

Censored Samples from a Normal Population

The estimation of the parameters of a censored sample taken from a normal population has been considered by many authors who have used different methods. In addition to the method of least squares and the method of maximum likelihood many other methods were used to obtain simple but highly efficient estimators. Halperin (1952) proved under mild regularity conditions that the ML-estimator of a single parameter from singly Type II censored samples is consistent, asymptotically normally distributed, and efficient. Cohen (1950) used the method of maximum likelihood to estimate the parameters of normal populations from singly and doubly censored samples for Type I censoring. In four later papers Cohen (1955, 1957, 1959, 1961) extended the results given in his 1950 paper. Gupta (1952) found maximum likelihood equations for estimators of the parameters of a normal population from singly Type II censored samples. Cohen and Gupta also give formulas for the asymptotic variances and covariances.

Lloyd (1952) applied the theory of least squares estimation to an ordered sample from distributions depending on location and scale parameters only. Gupta (1952) derived the best linear estimators for singly censored samples ($n \leq 10$) from normal populations. Sarhan and Greenberg (1956, 1958, 1962) estimated the mean and standard deviation of normal populations from singly and doubly censored samples ($n \leq 20$) by the method of least squares. Tables for the coefficients of these best unbiased linear estimators (BLU) as well as the variance and covariance are given by Sarhan and Greenberg (1962).

Gupta (1952) derived alternative linear estimators which are almost as efficient as the BLU estimators for $10 \leq n \leq 20$. The results were extended by Prescott (1970) to $n \leq 25$. These estimators depend only on the expected values of normal order statistics. Tables of these order statistics are given by Harter (1961). Plackett (1958) showed that maximum-likelihood estimators are asymptotically linear and that the best linear unbiased estimators are asymptotically normal and efficient. Harter and Moore (1966) and Harter (1970) consider the ML-estimators for Type II censoring and perform a simulation study which reveals that the ML-estimators have mean squared errors (MSE) which are smaller than the variances of the BLU-estimators for $n \geq 10$.

Since the likelihood equations are solvable only by iterative methods and the linear estimators involve the use of extensive tables, many authors considered simplified estimators. Saw (1959) developed simplified unbiased estimators of the mean and variance given a singly censored sample from a normal population ($n \leq 20$), but tables still have to be used. Dixon (1957, 1960) also gives simplified estimators ($n \leq 20$) which necessitate the use of tables. These estimators for the mean and standard deviation are almost as efficient as the BLU-estimators. Plackett (1958) derived a linearized maximum likelihood estimator for the standard deviation of a normal population from censored samples ($n \leq 10$). For this estimator tables have to be used. Tiku (1967) considers the ML-estimators and describes a method to approximate the likelihood equations such that the computation effort to determine ML-estimators is reduced substantially. After solving very simple equations explicit formulas for the estimators are derived. To facilitate the application of his method tables were prepared by Tiku (1970). Healy (1978) derived an estimator for the standard deviation suitable for symmetrically censored normal samples. Prescott (1979) extends Healy's estimator to asymmetric censoring. These estimators also make use of tables prepared by the authors. Explicit estimators for the mean and standard deviation of a normal population based on singly Type I censored samples are derived by Perrson and Rootzen (1977). No tables have to be used. The method used by Perrson and Rootzen was extended to doubly censored samples of Type I and Type II by Schneider (1984). These estimators are almost as efficient as the ML-estimators. Chernoff, Gastwirth, and Johns (1967) derived linear asymptotically efficient estimators for singly and doubly Type II censored samples from distributions dependent on location and scale parameters only. But the efficiency of these estimators for small sample sizes compared with the ML-estimators depends on the symmetry of the censoring and is very poor for singly censored samples.

In Section 4.1 order statistics and multiple censoring are considered. Section 4.2 deals with the maximum likelihood estimators. Section 4.3 considers the method of least squares to obtain linear estimators. Section 4.4 presents various simplified estimators. Bias correction methods are given in Section 4.5. A comparative study of the estimators is given in Section 4.7. Optimal spacing is reviewed in Section 4.6.

4.1 ORDER STATISTICS AND MULTIPLE CENSORING

In this section we shall give some general results concerning order statistics and general censoring. We consider a random sample X_1, X_2, ..., X_n, drawn from a normal population with mean μ and standard deviation σ. Rearranging them in the order of their magnitude yields the order statistics

$$X_{(1)} \leqslant X_{(2)} \leqslant \cdots \leqslant X_{(n)}$$

where $X_{(1)}$ is the smallest value and $X_{(n)}$ the largest, respectively. $X_{(i)}$ is called the ith order statistic. If we select $j(1 \leqslant j \leqslant n)$ integers such that

$$1 \leqslant n_1 \leqslant n_2 \leqslant \cdots \leqslant n_j \leqslant n$$

the joint pdf of the j order statistics $X_{(n_1)}, X_{(n_2)}, \ldots, X_{(n_j)}$ is given by

$$f(x_{(n_1)}, \ldots, x_{(n_j)}; \mu, \sigma) = \begin{cases} \dfrac{n!}{\prod_{i=1}^{j+1} (n_i - n_{i-1} - 1)!} \prod_{i=1}^{j+1} \left\{ \Phi\left[\dfrac{x_{(n_i)} - \mu}{\sigma}\right] - \Phi\left[\dfrac{x_{(n_{i-1})} - \mu}{\sigma}\right]\right\}^{n_i - n_{i-1} - 1} \\ \times \prod_{i=1}^{j} \phi(x_{(n_i)}) \quad \text{for } x_{(n_1)} < \cdots < x_{(n_j)} \\ 0 \quad \text{elsewhere} \end{cases} \tag{4.1}$$

where $n_0 = 0$, $n_{j+1} = n + 1$ and $x_{(n_0)} = -\infty$, $x_{(n_{j+1})} = +\infty$ (Ogawa, 1951). The pdf of the ith order statistics is then

$$f(x_{(i)}; \mu, \sigma) = \frac{n!}{(i-1)!(n-1)!} \Phi^{i-1}\left(\frac{x_{(i)} - \mu}{\sigma}\right)$$

$$\times \left[1 - \Phi\left(\frac{x_{(i)} - \mu}{\sigma}\right)\right]^{n-i} \phi\left(\frac{x_{(i)} - \mu}{\sigma}\right) \tag{4.2}$$

With the latter pdf we are able to calculate $E[x_{(i)}]$, $var[x_{(i)}]$, $E[X_{(i)}X_{(j)}]$ and $cov[X_{(i)}, X_{(j)}]$. Tables of the expected values of normal order statistics and the expected values of products of these order statistics up to $n = 20$ for $\mu = 0$, $\sigma = 1$ were first published by Teichroew (1956). Variances and covariances of the order statistics can be calculated from these tables and were published by Sarhan and Greenberg (1962). Harter (1970) published the expected values of the order statistics up to $n = 400$.

We are now going to describe the case of multiple censoring. Multiple Type I censoring can be defined by k intervals

$$(x_1, x^1), (x_2, x^2), \ldots, (x_k, x^k), \; -\infty \leq x_1 < x^1 < x_2 < \cdots < x^k \leq \infty$$

The observations which lie in the intervals are censored. If $x_1 = -\infty$ ($x^k = +\infty$), then the smallest (largest) observations are Type I censored. If $x_1 = -\infty$, $k = 1$ and $x^1 < \infty$ then the sample is singly Type I censored at the left and if $-\infty < x_1$, $k = 1$, $x^1 = \infty$ the sample is singly Type I censored at the right. We say a sample is doubly Type I censored if $-\infty = x_1$, $k = 2$, $x^2 = \infty$.

We notice that the information about the number of censored values might be different in various situations. We might know the numbers of censored observations in each interval ($r_1, \ldots, r_k$) or we might know only the sum $r = r_1 + \cdots + r_k$ of the censored values, or we might know, r_1, r_k and the sum $r_2 + \cdots + r_{k-1}$. Different information about the number of censored values yields different estimators for μ and σ. In a similar way we define multiple Type II censoring. In multiple Type II censoring the observations which lie within the sample percentile ranges defined by

$$(p_1, p^1), (p_2, p^2), \ldots, (p_k, p^k)$$

$$0 \leq p_1 < p^1 < \cdots < p_k < p^k \leq 1$$

are censored. If $p_1 = 0$ ($p^k = 1$) the smallest (largest) segment of observations is censored. As in the Type I censoring case we define left censoring ($p_1 = 0$, $p^1 < 1$, $k = 1$), right censoring ($0 < p_1$, $p^1 = 1$, $k = 1$) and double censoring ($p_1 = 0$, $p^2 = 1$, $k = 2$) of Type II. The number of censored observations r_i are now predetermined by p_i and p^i, but the intervals (x_i, x^i) are random.

For Type I as well as for Type II censoring the joint pdf has the form (4.1), where in the case of Type I censoring, $x_{(i)}$ is predetermined and n_i random, and in the case of Type II censoring, $x_{(i)}$ is random and n_i predetermined. The method of maximum likelihood might thus be applied to find estimators for μ and σ. In the following sections we will only deal with right, left, and double censoring. An exception is a problem which is denoted in literature as optimal spacing, which is described in Section 4.7.

The information matrix for multiple Type II censoring was derived by Chernoff et al. (1967). It turns out that this matrix is identical with the information matrix for partially grouped data, considered by Kulldorf (1961). To determine the information matrix, let

$$L_1(y) = -\phi'(y)/\phi(y) = y \tag{4.3}$$

$$L_2(y) = -[1 + y\phi'(y)/\phi(y)] = -1 + y^2 \tag{4.4}$$

$$\phi_i = \phi(\lambda_i), \quad \phi^i = \phi(\lambda^i) \tag{4.5}$$

Let E be the complement of the union of the intervals (λ_i, λ^i). The information matrix is then $\sigma^{-2}J$ where

$$J_{11} = \int_E L_1^2(y)\phi(y)\, dy + \sum_{i=1}^{k} (\phi^i - \phi_i)^2/(p^i - p_i) \tag{4.7}$$

$$J_{12} = J_{21} = \int_E L_1(y)L_2(y)\phi(y)\, dy + \sum_{i=1}^{k} (\phi^i - \phi_i)$$

$$\times\ (\lambda^i\phi^i - \lambda_i\phi_i)/(p^i - p_i) \tag{4.8}$$

$$J_{22} = \int_E L_2^2(y)\phi(y)\,dy + \sum_{i=1}^{k} (\lambda^i\phi^i - \lambda_i\phi_i)^2/(p^i - p_i) \tag{4.9}$$

The inverse of the information matrix corresponds then to the Cramer-Rao bound for multiple Type II censoring.

The ML-estimators of μ and σ are the same whether Type I or Type II censoring is used; we only have to replace the censoring points $(x_{(i)}, x^{(i)})$ (in Type I censoring) with the realizations of the corresponding random variables in Type II censoring. The information matrix of Type I and Type II censoring are in principle different. But the modifications, which have to be made to obtain the information matrix for Type I censoring, are obvious. One has to substitute $\Phi(\lambda_i)$ and $\Phi(\lambda^i)$ for p_i and p^i, respectively.

4.2 MAXIMUM LIKELIHOOD ESTIMATORS

In this section we derive the maximum likelihood estimators (MLE) for the mean μ and the standard deviation σ. Since there are some important practical applications, for instance life tests, where only singly censored samples appear, we shall deal with the singly censored case separately although it can be treated as a special case of double censoring. Furthermore, the solution procedures become simpler in the singly censored case as we shall see below. This section is based on the results of Cohen (1950, 1955, 1957, 1959, 1961), Gupta (1952), and Harter and Moore (1966).

Singly Censored Samples

The maximum likelihood estimators for Type I and Type II censoring differ only in the interpretation of the censoring points $x_r(x_\ell)$ and the number of censored observations $n_r(n_\ell)$. For Type I censoring, x_r and x_ℓ are fixed and the number of censored observations n_r and n_ℓ are random. For Type II censoring it is the inverse, that is, n_r and n_ℓ are fixed while x_r and x_ℓ are random. The variance of the estimators in the two types of censoring are in principle different but this difference diminishes with increasing sample size.

We shall first derive the maximum likelihood estimators for Type I censoring. Then the obvious changes will be made to obtain the maximum likelihood estimators for the Type II censoring.

Type I censored samples. Consider a random sample of size n, $(X_1, \ldots, X_n)$, taken from a normal population with mean μ and standard deviation σ. Only those n_0 observations x_i which lie numerically above a pre-assigned value $x_\ell \leq x_i$, $i = 1, \ldots, n_0$ (left censoring of Type I) or below a pre-assigned value $x_i \leq x_r$, $i = 1, \ldots, n_0$ (right censoring of Type I) are available. In contrast to a truncated normal distribution, the number $n - n_0$ of not observed values is known after the sample is taken.

When n is finite there is a non-zero probability that $n_0 = 0$, so that no observations are available, and in this case it is not possible to give a point estimate for μ and σ; therefore, a procedure has to be adopted for such an instance. But we must notice that any procedure will affect the bias and variance of the estimators.

We note that there is a symmetry such that we only have to derive estimators for right censoring; left censored samples can then be treated as right censored when we multiply the x_i values by -1.

Let $(x_1, \ldots, x_{n_0})$ be the observations which are available from the sample of size n; furthermore, let $n_r = n - n_0$ be the number of unobserved values. The pdf of the population from which the sample of size n is drawn is assumed to be normal with mean μ and variance σ^2. Thus, the likelihood function is

$$L(\mu, \sigma) \propto \sigma^{-n_0} \left[1 - \Phi\left(\frac{x_r - \mu}{\sigma}\right)\right]^{n_r} \exp\left[-\frac{1}{2}\sum_{i=1}^{n_0} (x_i - \mu)^2/\sigma^2\right] \tag{4.10}$$

The natural logarithm of the likelihood function is up to a constant

$$\log L(\mu, \sigma) = n_r \log [1 - \Phi(u_r)] - n_0 \log \sigma - \frac{1}{2}\sum_{i=1}^{n_0} (x_i - \mu)^2/\sigma^2 \tag{4.11}$$

where $u_r = (x_r - \mu)/\sigma$. The likelihood equations are then

$$\frac{\partial \log L}{\partial \mu} = n_r \frac{\phi(u_r)}{1 - \Phi(u_r)} \frac{1}{\sigma} + \sum_{i=1}^{n_0} (x_i - \mu)/\sigma^2 = 0 \tag{4.12}$$

and

$$\frac{\partial \log L}{\partial \sigma} = n_r \frac{\phi(u_r)}{1 - \Phi(u_r)} \frac{u_r}{\sigma} - \frac{n_0}{\sigma} + \sum_{i=1}^{n_0} (x_i - \mu)^2/\sigma^3 = 0 \tag{4.13}$$

Let $h = n_r/n$ and $W(u) = \phi(u)/\Phi(u)$ as defined in Chapter 2; we rewrite the likelihood equations in order to obtain Cohen's (1959) estimators

$$(\mu - \bar{x})/\sigma - [h/(1 - h)]\widetilde{W}(u_r) = 0 \tag{4.14}$$

$$s^2 + (\bar{x} - \mu)^2 - \sigma^2\{1 - u_r[h/(1 - h)]\widetilde{W}(\tilde{u}_r)\} = 0 \tag{4.15}$$

where

$$\bar{x} = \sum_{i=1}^{n_0} x_i/n_0$$

$$s^2 = \sum_{i=1}^{n_0} (x_i - \bar{x})^2/n_0$$

are the mean and variance of the observed values respectively. We notice the similarity of these equations with the equations derived in Section 3.1 for the truncated case. Let

$$Y(h, u_r) = [h/(1 - h)]\widetilde{W}(u_r) \tag{4.16}$$

Then we obtain Cohen's (1959) ML-estimators

$$\sigma^2_{ML} = s^2 + \lambda(h, u_r)(x_r - \bar{x})^2 \tag{4.17}$$

$$\mu_{ML} = \bar{x} + \lambda(h, u_r)(x_r - \bar{x}) \tag{4.18}$$

where

$$\lambda(h, u_r) = \frac{Y(h, u_r)}{Y(h, u_r) + u_r} \tag{4.19}$$

and u_r is the solution to

$$\frac{1 - Y(h, u_r)[Y(h, u_r) + u_r]}{[Y(h, u_r) + u_r]^2} = \frac{s^2}{(x_r - \bar{x})^2} \tag{4.20}$$

$\underline{\sigma^2 \text{ known}}$. We first consider the situation where σ^2 is known. Let $\delta = (\mu_{ML} - \bar{x})/\sigma$ and $\theta = (x_r - \bar{x})/\sigma$, then from (4.14) we obtain the equation

$$\delta = Y(h, u_r)$$

or, using $u_r = \theta - \delta$,

$$\delta = Y(h, \theta - \delta) \tag{4.21}$$

A Taylor expansion yields

$$\delta = Y(h, \theta) - \delta Y'(h, \theta)$$

and thus

$$\delta = \frac{Y(h, \theta)}{1 + Y'(h, \theta)} \tag{4.22}$$

Finally we obtain the Taylor approximation for μ_{ML}

$$\mu_{ML} = \bar{x} + \sigma Y(h, \theta)/[1 + Y'(h, \theta)] \tag{4.23}$$

where

$$Y'(h, \theta) = \frac{h}{1 - h}\{\phi^2(\theta) - \theta\phi(\theta)[1 - \Phi(\theta)]\}/[1 - \Phi(\theta)]^2 \tag{4.24}$$

We noticed in Section 3.1 that the Taylor approximation is accurate enough for $\theta \geqslant 2$. Since the quotient $Y/(1 + Y')$ in the censored case depends on two parameters θ and h, it is not possible to use rational function approximation as in the truncated case.

$\underline{\sigma^2 \text{ unknown}}$. When the variance is unknown estimates are obtained by solving equation (4.20) using the Newton-Raphson iteration method.

A FORTRAN program is given in Appendix B. Tables and nomograms were given by Cohen (1959). More extensive tables were published by Cohen (1961). Cooley and Cohen (1970) and Schmee and Nelson (1977) give even more extensive tables in technical reports. All tables use Cohen's (1959) equation (4.20) in order to obtain the auxiliary estimation function (4.19). For convenience, we have prepared Table A2, which is given in Appendix A. This table is slightly more extensive than the table given by Cohen (1961). The entries give λ as a function of $\omega = s^2/(x - x_r)^2$ and $h = n_r/n$. The auxiliary function λ is then used to obtain maximum likelihood estimators via equations (4.17) and (4.18). Also, a nomograph for determining λ was given by Cohen (1959).

Asymptotic variance. The variance and covariance of the estimators μ_{ML} and σ_{ML} are obtained by inverting the Fisher information matrix

$$\mathrm{ASCV}(\mu_{ML}, \sigma_{ML}) = \frac{\sigma^2}{E[n_0]} \frac{1}{J_{11}J_{22} - J_{12}^2} \begin{bmatrix} J_{22} & -J_{12} \\ -J_{12} & J_{11} \end{bmatrix}$$

The elements are

$$J_{11} = 1 + W(u_r)[W(-u_r) - u_r] \tag{4.25}$$

$$J_{12} = J_{21} = -W(u_r)\{1 - u_r[W(-u_r) - u_r]\} \tag{4.26}$$

$$J_{22} = 2 + u_r J_{12} \tag{4.27}$$

and

$$E[n_0] = \Phi(u_r)n \tag{4.28}$$

For left censoring we replace u_r by $-u_\ell$. The joint ML-estimators tend to a multivariate normal distribution, with dispersion matrix given above.

Type II censored samples. Consider a random sample of size n, $(X_1, \ldots, X_n)$, taken from a normal population with mean μ and standard deviation σ. Only the n_0 smallest (right censoring of Type II) or the n_0 largest (left censoring of Type II) are available. The number n_0 is pre-assigned and thus is not a random variable as in Type I censoring. We consider the ordered samples

$$x_{(n_\ell+1)}, x_{(n_\ell+2)}, \ldots, x_{(n)} \qquad \text{(left censoring } n_\ell = n - n_0)$$

$$x_{(1)}, x_{(2)}, \ldots, x_{(n-n_r)} \qquad \text{(right censoring } n_r = n - n_0)$$

Due to the symmetry of the normal distribution, we again deal only with a right-censored sample. The likelihood function is

$$L(\mu, \sigma) \propto \sigma^{-n_0}\{1 - \Phi[(x_{(n-n_r)} - \mu)/\sigma]\}^{n_r} \exp\left[-1/2\sum_{i=1}^{n_0} (x_i - \mu)^2/\sigma^2\right] \tag{4.29}$$

where $n_0 = n - n_r$. The natural logarithm of the likelihood function is up to a constant

$$\log L(\mu, \sigma) = n_r \log[1 - \Phi(\tilde{u}_r)] - n_0 \log \sigma - 1/2\sum_{i=1}^{n_0}(x_i - \mu)^2/\sigma^2 \tag{4.30}$$

where $\tilde{u}_r = (x_{(n-n_r)} - \mu)/\sigma$. Thus, setting $x_r = x_{(n-n_r)}$, we obtain the same likelihood equations as in the Type I censored case. We only have to keep in mind that for Type II censoring $x_{(n-n_r)}$ is a random variable and n_r is fixed, while in the Type I censored case n_r is a random variable and x_r is fixed. But we are able to use exactly the same estimation procedure as described above.

Asymptotic variance. The variance and covariance of the estimators μ_{ML} and σ_{ML} are obtained by inverting the Fisher information matrix:

$$\text{ASCV}(\mu_{ML}, \sigma_{ML}) = \frac{\sigma^2}{n_0} \frac{1}{J_{11}J_{22} - J_{12}^2} \begin{bmatrix} J_{22} & -J_{12} \\ -J_{12} & J_{11} \end{bmatrix}$$

The elements are

$$J_{11} = 1 + \frac{n_r}{n_0} W(-u_r)[W(-u_r) - u_r] \tag{4.31}$$

$$J_{12} = J_{21} = -\frac{n_r}{n_0} W(-u_r)\{1 - u_r[W(-u_r) - u_r]\} \tag{4.32}$$

$$J_{22} = 2 + u_r J_{12} \tag{4.33}$$

Notice that the covariance matrix for Type I and Type II censoring are different. But asymptotically they are equivalent since $n_r/n_0 W(-u_r) \longrightarrow W(u_r)$ as $n \longrightarrow \infty$.

If we have a left censored sample of Type II we substitute $-u_\ell$ for u_r.

In order to permit ready evaluation of the asymptotic variances and covariances of maximum likelihood estimators tables were prepared by Bliss (1937), Gupta (1952), Hald (1949), Cohen and Woodward (1953) and Cohen (1961). For convenience we added Table A3, which gives the variance factors for singly truncated and singly Type I or Type II censored samples, in Appendix A.

To illustrate the ease with which the tables can be used, we select three examples that have been previously considered by various authors.

Example 4.1

Right censored Type II sample. Data for this sample were given by Cohen (1959, 1961). The example was originally considered by Gupta (1952). A sample of size n = 300 is taken and the 181 (= n_r) largest observations were Type II censored. The sample mean and sample variance are $\bar{x} = 1{,}304.832$, and $s^2 = 12{,}128.250$, respectively. The largest observation was $x_{(119)} = 1{,}450.000$. We calculate $\omega = s^2/(\bar{x} - x_{(119)})^2 = 0.575515$ and $h = 181/300 = 0.6033$. Two-way linear interpolation in Table A2 of Appendix A yields $\lambda = 1.36$. Using (4.17) and (4.18), we then compute $\hat{\mu} = 1{,}502$, $\hat{\sigma}^2 = 40{,}789$ and $\hat{\sigma} = 202$. For the variance and covariance we enter Table 3 of Appendix A with h = 60.33% and interpolate linearly to obtain $\gamma_{11} = 2.022$, $\gamma_{12} = 1.051$, and $\gamma_{22} = 1.635$. With these values the approximate variances and covariances of the estimators can be calculated. But note that confidence intervals based on these variances are usually not reliable since the estimators are correlated even asymptotically. Asymptotic correct confidence intervals are considered in Chapter 7.

We also notice that the maximum likelihood estimators are biased. A bias correction is discussed in Section 4.5.

Example 4.2

Gupta (1952) gives the days on which the first 7 of a sample of 10 mice died after being inoculated with a culture of tuberculosis. The times of death were 41, 44, 46, 54, 55, 58, 60 days after inoculation. Assuming lognormal distribution of the survival time we take the logarithm with base 10 of the data. The sample mean and variance of the logs of the data is $\bar{x} = 1.7048$ and $s^2 = 0.003547$. The fraction censored is $h = 0.3$ and $\omega = s^2/(\bar{x} - x_{(7)}) = 0.659$. Linear interpolation in Table 2 of Appendix A yields $\lambda = 0.5125$. We then obtain the maximum likelihood estimates $\hat{\mu} = 1.742$, $\hat{\sigma}^2 = 0.006309$, and $\hat{\sigma} = 0.07943$. For the variance and covariance we enter Table A3 of Appendix A with $h = 0.3$. We obtain $\gamma_{11} = 1.138$, $\gamma_{22} = 0.8197$, and $\gamma_{12} = 0.2066$. With these values we can compute the approximate variances and covariance as $\text{var}(\hat{\mu}) = 0.1138\hat{\sigma}^2$, $\text{var}(\hat{\sigma}) = 0.08197\hat{\sigma}^2$, and $\text{cov}(\mu, \sigma) = 0.02066\hat{\sigma}^2$. Notice that the estimates are biased.

Example 4.3

The following locomotive control example was given by Nelson (1982). Table 4.1 shows singly Type I censored mileage data with $n_0 = 37$ failures in a sample of $n = 96$ locomotive controls. Assuming lognormal distribution of the mileage to failure it was desired to find estimates of the parameters of the lognormal distribution. Working with the base 10 logs of the data; the sample mean and variance are $\bar{x} = 1.921$ and $s^2 = 0.0307$, respectively. The fraction censored is $h = (96 - 37)/96 = 0.615$, and $\omega = s^2/(\bar{x} - x_r) = 0.698$. Linear two-way interpolation in Table A2 of Appendix A yields $\lambda = 1.439$. Then the MLEs are $\hat{\mu} = 2.222$ and $\sigma = 0.306$. We again notice that the MLEs are biased; but since the sample size is fairly large we can expect that the bias is small.

Doubly Censored Samples

Type I censored samples (n_r and n_ℓ known). Consider a random sample of size n, $X_1, \ldots, X_n$, taken from a normal population with mean μ and standard deviation σ. Only those n_0 observations x_i which lie numerically between the preassigned values x_ℓ and x_r ($x_\ell < x_r$) are available. Furthermore, it is assumed that the number of observations n_ℓ falling below x_ℓ and the number of observations n_r falling

TABLE 4.1 Thousands of Miles to Failure for Locomotive Controls

22.5	37.5	46.0	48.5	51.5	53.0	54.5	57.5	66.5	68.0	69.5	76.5
77.0	78.5	80.0	81.5	82.0	83.0	84.0	91.5	93.5	102.5	107.0	108.5
112.5	113.5	116.0	117.0	118.5	119.0	120.0	122.5	123.0	127.5	131.0	132.5
134.0											

Note: 59 controls operated for 135.0 thousand miles without failure.
Source: From Nelson (1982).

above x_r are known after the sample is drawn, i.e., n_ℓ and n_r are random. Let $n_0 = n - n_\ell - n_r$, and let us further assume $n_0 > 0$, and $x_1, \ldots, x_{n_0}$ are the observations which are available; then the likelihood function is

$$L(\mu, \sigma) \propto \sigma^{-n_0} \Phi\left(\frac{x_\ell - \mu}{\sigma}\right)^{n_\ell} \left[1 - \Phi\left(\frac{x_r - \mu}{\sigma}\right)\right]^{n_r} \times \exp\left[-1/2 \sum_{i=1}^{n_0} (x_i - \mu)^2/\sigma^2\right] \tag{4.34}$$

The natural logarithm of the likelihood function is up to a constant

$$\log L(\mu, \sigma) = n_\ell \log \Phi(u_\ell) + n_r \log[1 - \Phi(u_r)] - n_0 \log \sigma - 1/2 \sum_{i=1}^{n_0} (x_i - \mu)^2/\sigma^2 \tag{4.35}$$

where $u_\ell = (x_\ell - \mu)/\sigma$ and $u_r = (x_r - \mu)/\sigma$. The likelihood equations are then

$$\frac{\partial \log L}{\partial \mu} = -\frac{n_\ell}{\sigma}\frac{\phi(u_\ell)}{\Phi(u_\ell)} + \frac{n_r}{\sigma}\frac{\phi(u_r)}{1 - \Phi(u_r)} + \sum_{i=1}^{n_0} (x_i - \mu)/\sigma^2 = 0 \tag{4.36}$$

$$\frac{\partial \log L}{\partial \sigma} = -\frac{n_\ell}{\sigma} u_\ell \frac{\phi(u_\ell)}{\Phi(u_\ell)} + \frac{n_r}{\sigma} u_r \frac{\phi(u_r)}{1 - \Phi(u_r)} - \frac{n_0}{\sigma} + \sum_{i=1}^{n_0} (x_i - \mu)^2/\sigma^3 = 0 \tag{4.37}$$

We define the functions

$$Y_\ell(q_\ell, u_\ell) = \frac{n_\ell}{n_0}\,\frac{\phi(u_\ell)}{\Phi(u_\ell)} = Y_\ell \tag{4.38}$$

$$Y_r(q_r, u_r) = \frac{n_r}{n_0}\,\frac{\phi(u_r)}{1 - \Phi(u_r)} = Y_r \tag{4.39}$$

where $q_\ell = n_\ell/n_0$ and $q_r = n_r/n_0$. Notice that $Y_\ell(q_\ell, u_\ell) = W(u_\ell)n_\ell/n_0$ and $Y_r(q_r, u_r) = W(-u_r)n_r/n_0$.

$\underline{\sigma^2 \text{ known}}$. Let us first consider the case in which σ^2 is known. From $\partial \log L/\partial\mu = 0$ we obtain the equation

$$\delta = Y_r\left(q_r, \frac{x_r - \bar{x}}{\sigma} - \delta\right) - Y_\ell\left(q_\ell, \frac{x_\ell - \bar{x}}{\sigma} - \delta\right) \tag{4.40}$$

where $\delta = (\mu_{ML} - \bar{x})/\sigma$ and thus

$$\mu_{ML} = \bar{x} + \delta\sigma$$

is the maximum likelihood estimator for known σ^2. The Newton-Raphson method can be used to solve equation (4.40). For light censoring or nearly symmetrical censoring ($n_r \approx n_\ell$) the following Taylor approximation can be used. Let $\theta_r = (x_r - \bar{x})/\sigma$ and $\theta_\ell = (x_\ell - \bar{x})/\sigma$; then

$$\delta = \frac{Y_r - Y_\ell}{1 + Y_r' - Y_\ell'} \tag{4.41}$$

which results from a Taylor expansion and truncation after the linear term. The derivatives Y_r' and Y_ℓ' are

$$Y_\ell' = -n_\ell/n_0[\phi^2(\theta_\ell) + \theta_\ell\phi(\theta_\ell)\Phi(\theta_\ell)]/\Phi^2(\theta_\ell) \tag{4.42}$$

$$Y_r' = n_r/n_0\{\phi^2(\theta_r) - \theta_r\phi(\theta_r)[1 - \Phi(\theta_r)]\}/[1 - \Phi(\theta_r)]^2 \tag{4.43}$$

σ^2 unknown. When σ^2 is unknown the equations (4.36) and (4.37) have to be solved simultaneously; the Newton-Raphson method can be applied. The application of this method to these equations is straightforward and is discussed by Cohen (1957) and a FORTRAN program was written by Swan (1969b). Here we shall use a modification already introduced in Chapter 3.

Let

$$H = \frac{1}{u_r - u_\ell}, \quad G = \frac{u_r}{u_r - u_\ell}$$

$$d = x_r - x_\ell, \quad a = (x_r - \bar{x})/d, \quad b = s^2/d^2$$

then H and G are solutions to the equations

$$a = (Y_r - Y_\ell)H + G \tag{4.44}$$

$$a^2 + b = H^2 - y_\ell H + aG \tag{4.45}$$

and the ML-estimates are

$$\mu_{ML} = x_r - Gd \tag{4.46}$$

and

$$\sigma_{ML} = H(x_r - x_\ell) \tag{4.47}$$

respectively. The Newton-Raphson method can be applied to solve the equations (4.44) and (4.45). To obtain starting values we suggest using one of the simplified estimators derived in Section 4.4. These estimators provide values which are very close to the ML-estimates. A FORTRAN program which calculates estimates μ_{ML} and σ_{ML} based on H and G is presented in Appendix B.

A FORTRAN routine based on the EM algorithm of Demster et al. (1977) was presented by Wolnetz (1979). The programs are discussed in the Appendix.

Asymptotic variance. The variance and covariance of the estimators μ_{ML} and σ_{ML} are obtained by inverting the Fisher information matrix. Let Z_ℓ and Z_r be as defined in (2.5) and (2.6), then

$$\mathrm{ASCV}(\mu_{ML}, \sigma_{ML}) = \frac{\sigma^2}{E[n_0]} \frac{1}{J_{11}J_{22} - J_{12}^2} \begin{bmatrix} J_{22} & -J_{12} \\ -J_{12} & J_{11} \end{bmatrix}$$

The elements are

$$J_{11} = -\{Z_\ell [W(u_\ell) + u_\ell] + Z_r[W(-u_r) - u_r] + 1\} \tag{4.48}$$

$$J_{22} = -\{2 + Z_\ell u_\ell [1 + u_\ell (W(u_\ell) + u_\ell)] - Z_r u_r [1 - u_r(W(-u_r) - u_r)]\} \tag{4.49}$$

$$J_{12} = \{Z_\ell [1 + u_\ell (W(u_\ell) + u_\ell)] - Z_r[1 - u_r(W(-u_r) - u_r)]\} \tag{4.50}$$

$$E[n_0] = n[\Phi(u_r) - \Phi(u_\ell)] \tag{4.51}$$

Type I censored samples (only n_0 known). This case was originally discussed by Cohen (1950). Consider a random sample of size n, $X_1, \ldots, X_n$, taken from a normal population with mean μ and standard deviation σ. Only those n_0 observations x_i which lie numerically between the preassigned values x_ℓ and x_r ($x_\ell < x_r$) are available. We deal now with the case that only n_0 is known after the sample is taken. Thus we only know that $n - n_0$ observations have fallen below x_ℓ or above x_r, without knowing their values. We assume that $n_0 > 0$ and the observations which are available are numbered as $x_1, \ldots, x_{n_0}$; then the likelihood function is

$$L(\mu, \sigma) \propto \sigma^{-n_0} \{1 - \Phi[(x_r - \mu)/\sigma] + \Phi[(x_\ell - \mu)/\sigma]\}^{n-n_0}$$

$$\times \exp\left[-1/2 \sum_{i=1}^{n_0} (x_i - \mu)^2/\sigma^2\right] \tag{4.52}$$

The natural logarithm of the likelihood function is up to a constant

$$\log L(\mu, \sigma) = (n - n_0) \log[1 - \Phi(u_r) + \Phi(u_\ell)] - n_0 \log \sigma$$

$$- 1/2 \sum_{i=1}^{n_0} (x_i - \mu)^2/\sigma^2 \tag{4.53}$$

where $u_\ell = (x_\ell - \mu)/\sigma$ and $u_r = (x_r - \mu)/\sigma$. The likelihood equations are then

$$\frac{\partial \log L}{\partial \mu} = \frac{n - n_0}{\sigma} \frac{\phi(u_r) - \phi(u_\ell)}{1 - \Phi(u_r) + \Phi(u_\ell)} + \sum_{i=1}^{n_0} (x_i - \mu)/\sigma^2 = 0 \qquad (4.54)$$

$$\frac{\partial \log L}{\partial \mu} = \frac{n - n_0}{\sigma} \frac{\phi(u_r)u_r - \phi(u_\ell)u_\ell}{1 - \Phi(u_r) + \Phi(u_\ell)} - \frac{n_0}{\sigma} + \sum_{i=1}^{n_0} (x_i - \mu)^2/\sigma^3 = 0 \qquad (4.55)$$

We define

$$Q_\ell = \frac{n - n_0}{n_0} \frac{\phi(u_\ell)}{1 - \Phi(u_r) + \Phi(u_\ell)} \qquad (4.56)$$

and

$$Q_r = \frac{n - n_0}{n_0} \frac{\phi(u_r)}{1 - \Phi(u_r) + \Phi(u_\ell)} \qquad (4.57)$$

With the notation given above we obtain the maximum likelihood estimators

$$\mu_{ML} = x_r - Gd \qquad (4.58)$$

and

$$\sigma_{ML} = Hd \qquad (4.59)$$

where H and G are solutions to

$$a = (Q_r - Q_\ell)H + G \qquad (4.60)$$

$$a^2 + b = H^2 - Q_\ell H + aG \qquad (4.61)$$

These equations can be solved by the Newton-Raphson method.

Asymptotic variance. The asymptotic variance and covariance is given by the inverse of the Fisher information matrix

$$\mathrm{ASCV}(\mu_{ML}, \sigma_{ML}) = \frac{\sigma^2}{E[n_0]} \frac{1}{J_{11}J_{22} - J_{12}^2} \begin{bmatrix} J_{22} & -J_{12} \\ -J_{12} & J_{11} \end{bmatrix}$$

Let Z_ℓ and Z_r be as defined in (2.5) and (2.6),

$$\tilde{Z}_r = \frac{\phi(u_r)}{1 - \Phi(u_r) + \Phi(u_\ell)} \tag{4.62}$$

and

$$\tilde{Z}_\ell = \frac{\phi(u_\ell)}{1 - \Phi(u_r) + \Phi(u_\ell)} \tag{4.63}$$

then the elements are

$$J_{11} = 1 - Z_r u_r + Z_\ell u_\ell + (Z_r - Z_\ell)(\tilde{Z}_r - \tilde{Z}_\ell) \tag{4.64}$$

$$J_{22} = (Z_r u_r - Z_\ell u_\ell)(\tilde{Z}_r u_r - \tilde{Z}_\ell u_\ell) + u_\ell Z_\ell - u_r Z_r$$
$$- Z_r u_r^3 + Z_\ell u_\ell^3 + 2 \tag{4.65}$$

$$J_{12} = Z_\ell u_\ell^2 - Z_r u_r^2 + (Z_r - Z_\ell)(\tilde{Z}_r u_r - \tilde{Z}_\ell u_\ell - 1) \tag{4.66}$$

Type II censored samples. Consider a random sample of size n, $X_1, \ldots, X_n$, taken from a normal population with mean μ and standard deviation σ. The n_ℓ smallest and the n_r largest observation are not known. We consider the ordered sample $x_{(n_\ell+1)}, \ldots, x_{(n-n_r)}$. In contrast to Type I censoring, n_ℓ and n_r are fixed before a sample is drawn. We assume $n_0 > 0$. The likelihood function is

$$L(\mu, \sigma) \propto \sigma^{-n_0} \{\Phi[(x_{(n_\ell+1)} - \mu)/\sigma]\}^{n_\ell} \{1 - \Phi[(x_{(n-n_r)} - \mu)/\sigma]\}^{n_r}$$
$$\times \exp\left[- 1/2 \sum_{i=n_\ell+1}^{n-n_r} (x_{(i)} - \mu)^2/\sigma^2\right] \tag{4.68}$$

If we set $x_\ell = x_{(n_\ell+1)}$ and $x_r = x_{(n-n_r)}$ we see that the likelihood function of Type I and Type II coincide up to a constant factor. Hence the likelihood equations are identical for Type I and Type II censoring if we replace x_ℓ by $x_{(n_\ell+1)}$ and x_r by $x_{(n-n_r)}$, respectively. The same holds if only n_0 instead of n_r and n_ℓ is known. We notice that the variances and covariances of the estimators are different for Type I and Type II censoring. This is due to the fact that both types of censoring induce different random variables. For Type I censoring n_r and n_ℓ are random and x_ℓ and x_r are fixed, while for Type II censoring n_r and n_ℓ are fixed and x_ℓ and x_r are random. If n_ℓ and n_r are known then the elements of the covariance matrix

$$\text{ASCV}(\mu_{ML}, \sigma_{ML}) = \frac{\sigma^2}{n_0} \frac{1}{J_{11}J_{22} - J_{12}^2} \begin{bmatrix} J_{22} & -J_{12} \\ -J_{12} & J_{11} \end{bmatrix}$$

are

$$J_{11} = \frac{n_\ell}{n_0} W(u_\ell)[W(u_\ell) + u_\ell] + \frac{n_r}{n_0} W(-u_r)[W-u_r) - u_r] - 1 \tag{4.68}$$

$$J_{22} = 2 + \frac{n_\ell}{n_0} W(u_\ell)u_\ell\{1 + u_\ell[W(u_\ell) + u_\ell]\}$$

$$- \frac{n_r}{n_0} W(-u_r)u_r\{1 - u_r[W(-u_r) - u_r]\} \tag{4.69}$$

$$J_{12} = \frac{n_\ell}{n_0} W(u_\ell)\{1 + u_\ell[W(u_\ell) + u_\ell]\}$$

$$- \frac{n_r}{n_0} W(-u_r)\{1 - u_r[W(-u_r) - u_r]\} \tag{4.70}$$

If only n_0 is known then the elements of the covariance matrix are

$$J_{11} = 1 - \frac{n - n_0}{n_0} [\tilde{Z}_r u_r - \tilde{Z}_\ell u_\ell (\tilde{Z}_r - \tilde{Z}_\ell)^2] \tag{4.71}$$

$$J_{22} = 2 + \frac{n - n_0}{n_0} [(\tilde{Z}_\ell u_\ell - \tilde{Z}_r u_r) - \tilde{Z}_r u_r^3 + \tilde{Z}_\ell u_\ell^3$$

$$+ (\tilde{Z}_r u_r - \tilde{Z}_\ell u_\ell)^2] \tag{4.72}$$

$$J_{12} = \frac{n - n_0}{n_0} [\tilde{Z}_\ell u_\ell^2 - \tilde{Z}_r u_r^2 + (\tilde{Z}_r - \tilde{Z}_\ell)(\tilde{Z}_r u_r - \tilde{Z}_\ell u_\ell - 1)] \tag{4.73}$$

As noticed before the difference of the variance in Type I and Type II censoring diminishes as n tends to infinity.

4.3 LINEAR UNBIASED ESTIMATORS FOR TYPE II CENSORING

The maximum likelihood estimators discussed in Section 4.2 have well-known properties, namely that they are consistent, asymptotically efficient, and normally distributed. Their disadvantage is that the computational procedure is rather unattractive. For small sample sizes the bias of the ML-estimators is usually quite strong and therefore unbiased estimators are often preferred in the small sample case. But we notice that the mean square error of the ML-estimators is significantly smaller than the variance of the best linear unbiased estimators when sample sizes are larger than $n = 10$. This was shown by Harter and Moore (1966) (see Section 4.7).

Gupta (1952) and Sarhan and Greenberg (1956), (1958), and (1962) considered the best linear unbiased (BLU) estimators. These estimators are linear functions of the order statistics. Tables for the coefficients as well as for the variance and covariance were prepared by Sarhan and Greenberg up to sample size $n = 20$. Gupta (1952) introduced simpler linear unbiased estimators which have high efficiency compared with the BLU-estimators. These simple estimators only make use of the expectations of the order statistics. Gupta's alternative linear estimators, intended for the special case of singly censored samples, were extended to doubly censored samples by Prescott (1970). He also extended the tables of the coefficients for Gupta's estimator which already existed for $n \leq 20$, to $n \leq 25$. Dixon (1957, 1960) also suggested simple linear estimators of high efficiency. The variance and the coefficients of the BLUE are given in Table A5 and Table A4 of Appendix A, respectively.

Best Linear Unbiased Estimators

Consider a random sample of size n, $X_1, \ldots, X_n$, from a normal population which is Type II censored on the right and left. Singly censoring is treated as a special case of doubly censoring here and thus no specific discussion of that case is included. Let

$$x_{(n_\ell+1)} \leqslant \cdots \leqslant x_{(n-n_r)}$$

be the ordered sample where the n_ℓ smallest and the n_r largest observations are not available. Furthermore, $z_{i,n}$ denotes the expected value of the ith order statistics of a sample of size n from a population which has a standard normal distribution. Then the expected value of the ith order statistic of a random variable X which is normally distributed with mean μ and variance σ^2 is given by

$$E[X_{(i)}] = \mu + \sigma z_{i,n}$$

Let $\underline{V}$ denote the variance-covariance matrix of the standard normal order statistics; then the variance covariance matrix of the order statistics $X_{(n_\ell+1)} \leqslant \cdots \leqslant X_{(n-n_r)}$ is simply $\sigma^2\underline{V}$. We notice that the matrix V varies with the sample size n, but for convenience we will omit the index n. The elements of $\underline{V}$ are tabulated by Sarhan and Greenberg (1952) up to sample size n = 20. In matrix form the equation above can be written as

$$E[X] = \underline{B\theta} \tag{4.74}$$

where $X' = (X_{(n_\ell+1)}, \ldots, X_{(n-n_r)})$, $\theta' = (\mu, \sigma)$ and

$$\underline{B} = \begin{bmatrix} 1 & z_{n_\ell+1,n} \\ \vdots & \\ 1 & z_{n-n_r,n} \end{bmatrix} \tag{4.75}$$

The least square estimators are given by the minimization

$$\min_{\underline{\theta}}[(\underline{X} - \underline{B\theta})'V^{-1}(\underline{X} - \underline{B\theta})] \tag{4.76}$$

which yields the solution (Gupta, 1952)

$$\underline{\theta}_{BLU} = (\underline{B}'\underline{V}^{-1}\underline{B})^{-1}\underline{B}'\underline{V}^{-1}\underline{X} \tag{4.77}$$

It is easily shown that the estimators θ_{BLU} are unbiased. The variance-covariance matrix is given by (Gupta, 1952)

$$\text{cov}\,(\underline{\theta}_{BLU}) = \sigma^2(\underline{B}'\underline{V}^{-1}\underline{B})^{-1} \tag{4.78}$$

Equation (4.77) shows that the estimators μ_{BLU} and σ_{BLU} are linear functions of the order statistics and might thus be written as

$$\mu_{BLU} = \sum_{i=n_\ell+1}^{n-n_r} a_{i,n} x_{(i)} \tag{4.79}$$

$$\sigma_{BLU} = \sum_{i=n_\ell+1}^{n-n_r} b_{i,n} x_{(i)} \tag{4.80}$$

The coefficients $a_{i,n}$ and $b_{i,n}$ only depend on the expectations and variance-covariance matrix of the order statistics from a standard normal distribution. They are tabulated by Sarhan and Greenberg (1962) for sample sizes up to n = 20. For convenience they are also given in Tables A4 and A5 in Appendix A, for samples sizes up to n = 10.

It was demonstrated by Plackett (1958) that the BLU-estimators are asymptotically efficient and normally distributed. They are thus asymptotically equivalent to the ML-estimators.

Gupta's Estimators

Gupta (1952) suggested alternative linear estimators for the mean and standard deviation of a normal population. His approach ignores the covariance of the order statistics. Consequently he substitutes the identity matrix for V and thereby obtains the estimator

$$\underline{\theta}_{LU} = (\underline{B}'\underline{B})^{-1}\underline{B}'\underline{X} \tag{4.81}$$

which is unbiased and has the variance-covariance matrix

$$\operatorname{cov}(\underline{\theta}_{LU}) = \sigma^2(\underline{B}'\underline{B})^{-1}\underline{B}'\underline{V}\underline{B}(\underline{B}'\underline{B})^{-1}$$

The estimators μ_{LU} and σ_{LU} are linear functions of the order statistics

$$\mu_{LU} = \sum_{i=n_\ell+1}^{n-n_r} a_i' x_{(i)} \tag{4.82}$$

$$\sigma_{LU} = \sum_{i=n_\ell+1}^{n-n_r} b_i' x_{(i)} \tag{4.83}$$

where the coefficients can be written as [Gupta (1952)]

$$a_i' = \frac{1}{n - n_\ell - n_r} - \frac{\bar{z}_n(z_{i,n} - \bar{z}_n)}{\sum_{j=n_\ell+1}^{n-n_r}(z_{j,n} - \bar{z}_n)^2}$$

$$b_i' = \frac{z_{i,n} - \bar{z}_n}{\sum_{j=n_\ell+1}^{n-n_r}(z_{j,n} - \bar{z}_n)^2} \qquad i = n_\ell + 1, \ldots, n - n_r$$

where

$$\bar{z}_n = \frac{1}{n - n_\ell - n_r} \sum_{i=n_\ell+1}^{n-n_r} z_{i,n} \tag{4.86}$$

The coefficients a_i' and b_i' are easy to compute and depend only on the expected values of the order statistics from a standard normal distribution. The latter ones are tabulated by Harter (1970) for sample sizes up to $n = 400$. The coefficients required for the alternative estimators are thus readily available for all practical sample sizes.

The variance of the estimators, however, involves the variance covariance matrix of the standard normal order statistics which is only available for sample sizes up to $n = 20$. Approximations suggested by Prescott (1970) may be used to compute the variance of the alternative estimators for sample sizes larger than $n = 20$.

It was shown by Ali and Chan (1964) that Gupta's alternative estimators are asymptotically efficient and normally distributed.

Dixon's Estimators

Dixon considers symmetrical and singly censored samples. When the sample is unsymmetrically censored he suggests ignoring some of the observations to obtain a symmetrically censored sample. In fact the efficiencies of estimators of the mean are in general only slightly affected by this procedure, provided of course that the censoring does not tend to 100%. For example, consider the case where $n_\ell = 3$, $n_r = 4$ and $n = 10$. The variance of the best linear estimator μ_{BLU} is $0.134\sigma^2$. When the fourth smallest value is omitted, i.e., $n_\ell = 4$, the variance of μ_{BLU} increases to $0.138\sigma^2$, which is an increase of less than 3%. See Section 4.6 as well for a more detailed discussion of the efficiency of this procedure.

Symmetrical censoring. For symmetrical censoring Dixon suggests the winsorized estimator (note $n_\ell = n_r$)

$$\mu_W = \frac{1}{n}[(n_\ell + 1)x_{(n_\ell+1)} + x_{(n_\ell+2)} + \cdots + x_{(n-n_\ell-1)} + (n_\ell + 1)x_{(n-n_\ell)}] \tag{4.87}$$

The obvious advantage of this estimator is that it is very easy to compute and does not involve the use of tables. Furthermore, the efficiency of this simple estimator is very high compared with that of the best linear estimator.

Singly censored samples. For singly censored samples Dixon suggests using the simple estimator

$$\mu_a = \frac{1}{n + a - 1}[ax_{(1)} + x_{(2)} + \cdots + x_{(n-n_r-1)} + (n_r + 1)x_{(n-n_r)}] \tag{4.88}$$

TABLE 4.2 Values of the Constant a for Dixon's Estimator

$n \backslash n_r$	1	2	3	4	5	6
3	0.0					
4	0.289	-0.866				
5	0.426	-0.426	-1.703			
6	0.506	-0.188	-1.143	-2.532		
7	0.560	-0.038	-0.821	-1.864	-3.361	
8	0.599	0.066	-0.609	-1.466	-2.591	-4.190
9	0.628	0.143	-0.458	-1.198	-2.122	-3.324
10	0.651	0.202	-0.345	-1.002	-1.800	-2.787
11	0.669	0.249	-0.255	-0.853	-1.562	-2.412
12	0.685	0.288	-0.183	-0.734	-1.378	-2.133
13	0.698	0.321	-0.123	-0.638	-1.231	-1.916
14	0.709	0.349	-0.073	-0.557	-1.109	-1.740
15	0.719	0.373	-0.029	-0.488	-1.008	-1.596
16	0.727	0.394	0.008	-0.429	-0.921	-1.473
17	0.735	0.413	0.041	-0.378	-0.846	-1.369
18	0.742	0.429	0.070	-0.332	-0.781	-1.278
19	0.748	0.444	0.097	-0.292	-0.723	-1.198
20	0.754	0.458	0.120	-0.256	-0.671	-1.127

Source: From Dixon (1960).

where "a" is a constant such that μ_a is an unbiased estimator of μ. Notice that a left censored sample can be treated as a right censored sample by changing the sign of the observations. In the following table we give the values of the constant "a" for sample sizes $n \leq 20$ and censoring $n_r \leq 6$. The entries of Table 4.2 are easily computed by setting the expected value of (4.88) equal to zero and using a table of expected order statistics. Dixon (1960) computed the constant "a" for $n \leq 20$ and $n_r \leq 6$.

Estimator of σ. Dixon (1957) pointed out (see also Section 4.7 for more detailed information) that the efficiency of the best linear estimator of σ drastically decreases even for light censoring. If, for example, only one of 10 observations is missing at one end of the ordered sample, the efficiency drops to 85%. The efficiency here is the ratio of the variances for censored and uncensored samples. A further slight loss of efficiency due to choosing a simplified estimator might thus be negligible. Dixon suggests therefore using only one or two ranges to obtain an estimator for σ which has still high efficiency compared with the best linear estimator.

For singly censored samples, for instance, an estimator of the standard deviation based on an optimum choice of one or two ranges has minimum efficiency 0.937 compared to BLU for sample sizes $n < 20$ and less than seven observations censored. When the order statistics, to be used for the estimation of σ, are selected, an appropriate multiplier must be used to give an unbiased estimate of the standard deviation. This multiplier is easy to compute by using the expected values of the order statistics.

For $n \leqslant 5$ Dixon suggests using the range

$$\sigma_b = b[x_{(n_0)} - x_{(1)}] \tag{4.89}$$

for $6 \leqslant n \leqslant 10$ the best choice is

$$\sigma_b = b[2x_{(n_0)} - x_{(1)} - x_{(2)}] \tag{4.90}$$

and for $11 \leqslant n \leqslant 20$ the optimal selection of two ranges is given in Table 4.3.

Note that the largest and smallest observations always get at least the weight b and -b, respectively. The entries in the table give the order statistics $x_{(i)}$ which get an additional weight (-b). Note that if $i \leqslant n/2$ the weight b has to be multiplied by (-1). When the largest observation is mentioned in the table it thus gets an additional weight b, i.e., it is weighted by $2 \times b$.

The values of the multiplier b are given in Table 4.3 and Table 4.4. Dixon (1960) also considers doubly censored samples.

Example 4.4 (Dixon's estimator of σ)

As an example we consider the case that the sample size is $n = 17$ and the two largest observations are not available. From Table 4.4

TABLE 4.3 Factor b for Dixon's Estimator σ_b and Singly Censored Samples

$n\backslash n_0$	2	3	4	5	6	7	8	9
3	1.182							
4		0.754						
5		0.860	0.603					
6			0.433	0.313				
7			0.474	0.355	0.276			
8				0.387	0.310	0.251		
9				0.414	0.337	0.281	0.234	
10					0.359	0.304	0.260	0.220

Note: if $n \leq 5$ use (4.89), otherwise use (4.90).

and Table 4.5 we learn that the smallest observation obtains the weight -0.215 while the largest available observation is assigned the weight 0.215. Furthermore, the third and 14th largest observations obtain the weight -0.215 and 0.215, respectively. Thus Dixon's estimator of σ is given by $\sigma_b = 0.215\,(x_{(15)} + x_{(14)} - x_{(1)} - x_{(3)}]$. If the three largest observations were censored, then from Tables 4.4 and 4.5 we obtain the estimator $\sigma_b = 0.225\,(2x_{(14)} - x_{(1)} - x_{(3)})$.

Example 4.5

We consider Example 4.2. The log (basis 10) survival times are reproduced in Table 4.6.

BLU Estimators

To obtain the BLU estimators we enter Table A4 in Appendix A. The weights for sample size $n = 10$ and $n_r = 3$ are displayed in Table 4.6. Multiplication of the survival times with the weights and summation yields $\hat{\mu} = 1.746$ and $\hat{\sigma} = 0.09119$. The variances and covariances of the estimators can be found in Table A5 of Appendix A. We obtain $\text{var}(\hat{\mu}) = \hat{\sigma}^2 0.1167$, $\text{var}(\hat{\sigma}) = \hat{\sigma}^2 0.0989$, and $\text{cov}(\mu, \hat{\sigma}) = \hat{\sigma}^2 0.026$. We notice that the BLU estimate $\hat{\sigma}$ considerably differs from the MLE given in Example 4.2 as $\sigma_{ML} = 0.07943$. Since the MLE $\hat{\sigma}$ is biased

TABLE 4.4 Factor b for Dixon's Estimator and Singly Censored Samples

$n \backslash n_0$	6	7	8	9	10	11	12	13	14	15	16	17	18	19
11	0.378	0.323	0.280	0.244	0.225									
12		0.339	0.297	0.262	0.231	0.231								
13		0.353	0.354	0.304	0.268	0.237	0.206							
14			0.360	0.319	0.284	0.255	0.227	0.212						
15			0.373	0.332	0.298	0.269	0.243	0.218	0.205					
16				0.344	0.310	0.282	0.257	0.234	0.222	0.199				
17				0.354	0.321	0.293	0.268	0.246	0.225	0.215	0.193			
18					0.331	0.303	0.279	0.257	0.237	0.218	0.208	0.197		
19					0.340	0.312	0.288	0.267	0.248	0.230	0.212	0.203		
20						0.320	0.297	0.276	0.257	0.239	0.223	0.207	0.193	0.195

TABLE 4.5 Order Statistics for Dixon's Estimator

$n \backslash n_0$	6	7	8	9	10	11	12	13	14	15	16	17	18	19
11	(2,6)	(2,7)	(2,8)	(2,9)	(2,9)									
12	–	"	"	"	(2,10)	(3,10)								
13	–	"	(3,8)	(3,9)	(3,10)	(3,11)	(3,12)							
14	–	–	"	"	"	"	"	(3,12)						
15	–	–	"	"	"	"	"	(3,13)	(3,13)					
16	–	–	–	"	"	"	"	"	"	(3,14)				
17	–	–	–	"	"	"	"	"	(3,14)	"	(3,15)			
18	–	–	–	–	"	"	"	"	"	(3,15)	"	(3,15)		
19	–	–	–	–	"	"	"	"	"	"	(3,16)	(3,16)		
20	–	–	–	–	–	"	"	"	"	"	"	(3,17)	(3,17)	(3,17)

Source: From Dixon (1960).

TABLE 4.6 Log Survival Times of Mice Data and BLU Weights

Log survival time	1.613	1.643	1.663	1.732	1.740	1.763	1.778
Weights for MEAN	0.0244	0.0636	0.0818	0.0962	0.1089	0.1207	0.5045
Weights for STD	-0.3252	-0.1758	-0.1058	-0.0502	-0.0006	0.0468	0.6107

downwards we might explain the difference by the bias. But on the other hand the variance of the BLUE is slightly higher than the variance of the MLE. In fact the mean square error of the MLE is still smaller than the variance of the BLUE. This result generally holds for both, the mean and standard deviation, as was shown by Harter and Moore (1966) by means of a simulation study. We shall discuss the efficiency of the estimators in Section 4.7.

Dixon's Estimators

To obtain Dixon's estimator for the mean we use Eq. (4.88) and enter Table 4.2.

$$\hat{\mu} = [-0.345(1.613) + 1.643 + 1.663 + 1.732 + 1.740$$
$$+ 1.763 + 4(1.778)]/(10 - 0.345 - 1)$$
$$\hat{\mu} = 1.7443$$

To compute Dixon's estimator for the standard deviation we use Eq. (4.91) and enter Table 4.3.

$$\hat{\sigma} = 0.304(2\ 1.778 - 1.613 - 1.643) = 0.0912$$

We notice that Dixon's estimates are very close to the BLU estimates.

4.4 SIMPLIFIED ESTIMATORS

In this section we discuss simple estimators which can be evaluated without using extensive tables or iterative procedures. Most of these estimators have a high efficiency. For a detailed discussion of the efficiency of the estimators see Section 4.6.

Linear Asymptotic Efficient Estimators

Chernoff et al. (1967) introduced linear estimators which are asymptotically efficient and normally distributed. Their statistics have the form

$$T_n = \frac{1}{n}\sum_{j=1}^{n} L\left(\frac{j}{n+1}\right)X_{(j)} \tag{4.91}$$

where

$$X_{(1)} \leqslant X_{(2)} \leqslant \cdots \leqslant X_{(n)} \tag{4.92}$$

represents the ordered random sample and $L(\cdot)$ is a continuous function such that T_n is asymptotically efficient, i.e., the variance covariance matrix is given by the inverse of the Fisher information matrix. We consider a random sample from a normal population which is Type II doubly censored. Let n_ℓ and n_r be the number of observations which are censored at the left and right respectively. Furthermore, let $p_\ell = n_\ell/n$ and $p_r = n_r/n$, and u_ℓ and u_r be defined by $p_\ell = \Phi(u_\ell)$ and $1 - p_r = \Phi(u_r)$, respectively.

The general theory given by Chernoff et al. applies to the family of location and scale distributions. Here we will present their results for the normal distribution. The inverse of the Fisher information matrix J^{-1} for Type II censoring was given above. The elements of the information matrix can also be written in the form

$$J_{11} = 1 - p_r - p_\ell - [u_r\phi(u_r) - u_\ell\phi(u_\ell)] + \frac{\phi^2(u_r)}{p_r} + \frac{\phi^2(u_\ell)}{p_\ell} \tag{4.93}$$

$$J_{12} = -(1 + u_r^2)\phi(u_r) + (1 + u_\ell^2)\phi(u_\ell) + \frac{u_\ell\phi^2(u_\ell)}{p_r} + \frac{u_r\phi^2(u_r)}{p_r} \tag{4.94}$$

$$J_{22} = 2(1 - p_r - p_\ell) - u_r(1 + u_r^2)\phi(u_r) + u_\ell(1 + u_\ell^2)\phi(u_\ell) + \frac{u_\ell^2\phi^2(u_\ell)}{p_\ell} + \frac{u_r^2\phi^2(u_r)}{p_r} \tag{4.95}$$

We now consider the linear estimators μ_{CH} and σ_{CH} which are given by

$$(\mu_{CH}, \sigma_{CH}) = (T_{\mu n}, T_{\sigma n})J^{-1} \tag{4.96}$$

$T_{\mu n}$ and $T_{\sigma n}$ are defined by the weight function $L(\cdot)$ for the uncensored observations and additional weights for the extreme values [Chernoff et al. (1967)]

$$L_\mu(p) = \begin{cases} 1 & \text{if } y = \Phi^{-1}(p) \in [u_\ell, u_r] \\ 0 & \text{otherwise} \end{cases} \tag{4.97}$$

$$L_\sigma(p) = \begin{cases} 2y & \text{if } y = \Phi^{-1}(p) \in [u_\ell, u_r] \\ 0 & \text{otherwise} \end{cases} \tag{4.98}$$

The additional weights are

$$a_{\mu_2} = -u_r \phi(u_r) + \phi^2(u_r)/p_r \tag{4.99}$$

$$a_{\sigma_2} = \phi(u_r) - u_r^2 \phi(u_r) + u_r \phi^2(u_r)/p_r \tag{4.100}$$

$$a_{\mu_1} = u_\ell \phi(u_\ell) + \phi^2(u_\ell)/p_\ell \tag{4.101}$$

$$a_{\sigma_1} = -\phi(u_\ell) + u_\ell^2 \phi(u_\ell) + u_\ell \phi^2(u_\ell)/p_\ell \tag{4.102}$$

Hence we obtain

$$T_{\mu n} = \frac{1}{n} \sum_{j=n_\ell+1}^{n-n_r} x_{(j)} + a_{\mu_1} x_{(n_\ell+1)} + a_{\mu_2} x_{(n-n_r)} \tag{4.103}$$

$$T_{\sigma n} = \frac{1}{n} \sum_{j=n_\ell+1}^{n-n_r} 2\Phi^{-1}\left(\frac{j}{n+1}\right) x_{(j)} + a_{\sigma_1} x_{(n_\ell+1)} + a_{\sigma_2} x_{(n-n_r)} \tag{4.104}$$

The estimators can therefore be written as

$$\mu_{CH} = (J_{22}T_{\mu n} - J_{12}T_{\sigma n})/(J_{11}J_{22} - J_{12}^2) \tag{4.105}$$

$$\sigma_{CH} = (J_{11}T_{\sigma n} - J_{12}T_{\mu n})/(J_{11}J_{22} - J_{12}^2) \tag{4.106}$$

where J_{ij} are the elements of the information matrix.

The estimators are asymptotically efficient and normally distributed as shown by Chernoff et al. (1967).

We shall now consider the case in which σ or μ is known.

σ known. If σ is known we choose

$$T_{\mu n} = \frac{1}{n}\sum_{j=n_\ell+1}^{n-n_r} x_{(j)} + a_{\mu_1} x_{(n_\ell+1)} + a_{\mu_2} x_{(n-n_r)} \tag{4.107}$$

where

$$a_{\mu_1} = u_\ell \phi(u_\ell) + \phi^2(u_\ell)/p_\ell \tag{4.108}$$

$$a_{\mu_2} = -u_r \phi(u_r) + \phi^2(u_r)/p_r \tag{4.109}$$

Then

$$\mu_{CH} = (T_{\mu n} - J_{12}\sigma)/J_{11} \tag{4.110}$$

is the asymptotically efficient estimator.

μ known. If μ is known we choose

$$T_{\sigma n} = \frac{2}{n}\sum_{j=n_\ell+1}^{n-n_r} \Phi^{-1}\left(\frac{j}{n+1}\right)x_{(j)} + a_{\sigma_1} x_{(n_\ell+1)} + a_{\sigma_2} x_{(n-n_r)} \tag{4.111}$$

where

$$a_{\sigma_1} = -\phi(u_\ell) + u_\ell^2\phi(u_\ell) + u_\ell\phi^2(u_\ell)/p_\ell \tag{4.112}$$

$$a_{\sigma_2} = \phi(u_r) - u_r^2\phi(u_r) + u_r\phi^2(u_r)/p_r \tag{4.113}$$

Then

$$\sigma_{CH} = (T_{\sigma n} - J_{12}\mu)/J_{22} \tag{4.114}$$

is the asymptotically efficient estimator. When the censoring is symmetric, J_{12} is zero and the estimators μ_{CH} and σ_{CH} are independent. Therefore, in the case of symmetrical censoring we are able to use the estimators derived for the known μ and known σ case.

Example 4.6

We consider two examples of symmetrical censoring. If the upper and lower 100 p percent of the observations are censored, then the asymptotic optimum linear estimate of μ and σ is formed by applying the known σ and known μ case. Let $n_r = n_\ell = (pn)$; then $u_\ell = \Phi^{-1}(p)$ and $u_r = \Phi^{-1}(1 - p)$. We note that $J_{12} = 0$ in (4.110) and (4.114); thus Chernoff's estimators are given by

$$\mu_{CH} = a_\mu[x_{(n_\ell+1)} + x_{(n-n_r)}] + \sum_{i=n_\ell+1}^{n-n_r} x_i/(nJ_{11})$$

and

$$\sigma_{CH} = a_\sigma[x_{(n_\ell+1)} + x_{(n-n_r)}] + \sum_{i=n_\ell+1}^{n-n_r} \Phi^{-1}\left(\frac{i}{n+1}\right)x_{(i)}/(nJ_{22})$$

When $p = 0.05$ then $J_{11} = 0.986$ and $a_\mu = 0.0437$. When $p = 0.10$ then $J_{11} = 0.966$, and $a_\mu = 0.086$.

Mean Difference Estimator for the Standard Deviation

Healy (1978) considers a linear estimator for the standard deviation which is based on Downton's (1966) estimator for uncensored samples. Downton's estimator is given by

$$\sigma_b = \frac{1}{n(n-1)} \sum_{i=1}^{n} (2i - n - 1) \tag{4.115}$$

where $b = \sqrt{\pi}$. Healy's estimator was intended for the special case of symmetrical censoring of Type II. His work has been extended by Prescott (1979) to permit its use in the general case of unsymmetrical censoring.

Consider a random sample of size n from a normal population. The n_ℓ smallest and the n_r largest observations are not available. Let $n_0 = n - n_\ell - n_r$, $p_0 = n_0/n$, $p_\ell = n_\ell/n$; then Healy's estimator is defined by

$$\sigma_H = \frac{b(p_0, p_\ell)}{n_0(n_0 - 1/2)} \sum_{i=n_\ell+1}^{n-n_r} (2i - n_0 - 1)x_{(i)} \tag{4.116}$$

The factor b is chosen such that the estimator σ_H is unbiased. Unfortunately the correct value of b depends on n, n_ℓ and n_r and not only on p_0 and p_ℓ. For large n, however, this fact can be ignored. Hence the true value of b is substituted by its asymptotic value which satisfies the equation

$$1 = \frac{b(p_0, p_\ell)}{p_0^2} \int_{p_\ell}^{p_0+p_\ell} u[2\Phi(u) - p_0 - 2p_\ell]\phi(u)\, du \tag{4.117}$$

Integration gives the closed form

$$b(p_0, p_\ell) = p^2 \left\{ \frac{1}{\sqrt{\pi}} [\Phi(\sqrt{2}\, u_r) - \Phi(\sqrt{2}\, u_\ell)] - p_0[\phi(u_\ell) + \phi(u_r)] \right\}^{-1} \tag{4.118}$$

where $\Phi(u_\ell) = p_\ell$ and $\Phi(u_r) = p_0 + p_\ell$. Table 4.7 shows the factor $b(p_0, p_\ell)$ for some selected values of p_0 and p_ℓ.

TABLE 4.7 The Unbiased Factor $b(p_0, p_\ell)$

$p_0 \backslash p_\ell$	0	0.01	0.02	0.03	0.04	0.05	0.10	0.15	0.20	0.25
0.99	1.823	1.823								
0.98	1.862	1.878	1.862							
0.97	1.897	1.920	1.920	1.897						
0.96	1.929	1.958	1.965	1.958	1.929					
0.95	1.960	1.993	2.004	2.004	1.993	1.960				
0.94	1.989	2.026	2.041	2.046	2.041	2.026				
0.93	2.017	2.057	2.076	2.085	2.085	2.076				
0.92	2.044	2.088	2.110	2.122	2.126	2.122				
0.91	2.070	2.117	2.143	2.158	2.165	2.165				

0.90	2.096	2.146	2.174	2.192	2.202	2.206	2.096			
0.875	2.158	2.216	2.251	2.275	2.292	2.302	2.264			
0.85	2.218	2.284	2.325	2.355	2.377	2.393	2.393	2.218		
0.825	2.277	2.350	2.397	2.432	2.460	2.481	2.511	2.415		
0.80	2.334	2.415	2.468	2.509	2.542	2.568	2.624	2.568	2.334	
0.75	2.447	2.544	2.610	2.662	2.705	2.741	2.845	2.845	2.741	2.447
0.70	2.559	2.674	2.753	2.817	2.871	2.918	3.070	3.117	3.070	2.918
0.65	2.671	2.807	2.900	2.978	3.044	3.102	3.304	3.397	3.397	3.304
0.60	2.786	2.945	3.055	3.147	3.227	3.298	3.556	3.696	3.741	3.696
0.55	2.904	3.090	3.220	3.329	3.425	3.511	3.832	4.025	4.118	4.118
0.50	3.026	3.245	3.398	3.528	3.642	3.745	4.140	4.395	4.541	4.588

Source: Prescott (1979).

The estimator is not exactly unbiased for finite n, since the asymptotic value of b is used. Investigations show, however, that the bias is small (about 3%) provided that not more than half the sample is censored (see Prescott, 1979).

Saw's Estimator

Saw (1959, 1961a) considered estimators for Type I and Type II censored samples. Suppose a singly Type II censored sample is given. Let us assume n_r observations are censored at the right. Define $n_0 = n - n_r$ and $\overline{x}_r = \Sigma_{i=1}^{n_0 - 1} x_i/(n_0 - 1)$, then Saw's estimators are given by

$$\mu_s = \epsilon \overline{x}_r + (1 - \epsilon)x_r \tag{4.119}$$

$$\sigma_s^2 = \alpha \sum_{i=1}^{n_0-1} (x_i - x_{n_0})^2 + \beta\left[\sum_{i=1}^{n_0-1} (x_i - x_{n_0})\right]^2 \tag{4.120}$$

the ϵ being chosen to make μ_s unbiased and α and β being chosen to make σ^2 unbiased and of minimum variance. These factors α, β and ϵ are dependent on n and n_r and are tedious to calculate. Saw (1959) gives a table of ϵ for sample sizes up to n = 20. The obvious advantage of these estimators is that only a small number of weights have to be tabulated. Saw (1961) also considers Type I censoring. The following linear estimators were suggested:

$$\mu_{s'} = e_0(p_\ell, p_r) \frac{1}{n_0} \sum_{i=n_\ell+1}^{n-n_r} x_i + e_i(p_\ell, p_r)x_\ell + e_2(p_\ell, p_r)x_r \tag{4.121}$$

$$\sigma_{s'} = a_0(p_\ell, p_r) \frac{1}{n_0} \sum_{i=n_\ell+1}^{n-n_r} x_i + a_1(p_\ell, p_r)x_\ell + a_2(p_\ell, p_r)x_r \tag{4.122}$$

The factors a_0, a_1, a_2, e_0, e_1 and e_2 are chosen such that the estimators are unbiased to the order of $1/n$.

Tiku's Estimators

The major disadvantage of the maximum likelihood estimators is that there is no explicit solution for them. Simulation results, presented in Section 4.7, show, however, that the mean square error of the ML-estimators is in general smaller than the mean square error of all other known estimators, even for small sample sizes ($n \geq 10$). There are therefore good reasons for looking for approximations to the likelihood equations to reduce the computational effort in finding ML-estimators. Tiku (1967) suggested approximating the likelihood equations by linear functions in order to obtain an explicit solution. These estimators are known as modified maximum likelihood (MML) estimators, and are asymptotically equivalent to the maximum likelihood estimators. Tiku's (1967) estimators are based on a first order Taylor approximation to the hazard function $\tilde{W}(u)$:

$$\tilde{W}(u) = \tilde{W}(c) + (u - c)\tilde{W}'(c) \tag{4.123}$$

The problem is to find c_ℓ and c_r which are in the neighborhood of the unknown $u_\ell = (x_\ell - \mu)/\sigma$ and $u_r = (x_r - \mu)/\sigma$, respectively. Tiku (1967) suggested using

$$\tilde{W}'(c) = [\tilde{W}(c) - \tilde{W}(c + h)]/h \tag{4.124}$$

where, for instance, for left censored samples,

$$\Phi(c_\ell) = p_\ell - \sqrt{[p_\ell(1 - p_\ell)/n]} \tag{4.125}$$

and

$$\Phi(c_\ell + h) = p_\ell + \sqrt{[p_\ell(1 - p_\ell)/n]} \tag{4.126}$$

Similar equations are obtained for right censored samples. Tiku (1967) states that for reasonable large n the interval (c, c + h) is likely to cover the unknown u. But, on the other hand, for sample sizes larger than 10, Tiku (1970) suggested using the first derivative of $\tilde{W}(u)$ in Eq. (4.123):

$$\tilde{W}'(c) = \tilde{W}(c)[\tilde{W}(c) - c] \tag{4.127}$$

To determine c one uses the intuitive equations

$$\Phi(c_\ell) = n_\ell/n \tag{4.128}$$

for left, and

$$\Phi(c_r) = 1 - n_r/n \tag{4.129}$$

for right censored samples. The explicit form of the first derivative is given in Eq. (4.24). Substituting the Taylor approximation (4.123) for the hazard function in the likelihood equations (4.36) and (4.37), Tiku obtained closed forms for the estimates. Instead of using the likelihood equations we use the simpler transformed equations (4.44) and (4.45) to derive Tiku's modified maximum likelihood estimates. Let us first write (4.123) in the form

$$\tilde{W}(u) = \alpha + u\beta \tag{4.130}$$

where

$$\beta = \tilde{W}'(c) \tag{4.131}$$

and

$$\alpha = \tilde{W}(c) - \tilde{W}'(c)c \tag{4.132}$$

For a doubly censored sample one obtains then

$$\beta_\ell = \phi(c_\ell)[\phi(c_\ell)/p_\ell - c_\ell]/p_\ell$$

$$\alpha_\ell = \phi(c_\ell)/p_\ell - \beta_\ell c_\ell$$

$$\beta_r = \phi(c_r)[\phi(c_r)/p_r - c_r]/p_r$$

$$\alpha_r = \phi(c_r)/p_r - \phi_r c_r$$

where $p_\ell = n_\ell/n$, $p_r = n_r/n$ and c_ℓ and c_r are given by (4.128) and (4.129), respectively. Before proceeding we recall the notations

$$d = x_{(n-n_r)} - x_{(n_\ell+1)}, \quad a = [x_{(n-n_r)} - \bar{x})/d, \quad b = s^2/d^2$$

and

$$\bar{x} = \Sigma x_i/n_0, \quad s^2 = \Sigma(x_i - \bar{x})^2/n_0$$

Tiku's MMLEs are then given by

$$\mu_{MML} = x_{(n-n_r)} - Gd \tag{4.135}$$

and

$$\sigma_{MML} = Hd \tag{4.136}$$

where H and G are solutions to the equations

$$a = \left[\left(\alpha_r + \beta_r \frac{G}{H}\right)\frac{n_r}{n_0} - \left(\alpha_\ell + \beta_\ell \frac{G-1}{H}\right)\frac{n_\ell}{n_0}\right]H + G \tag{4.137}$$

$$b + a^2 = H^2 - \left(\alpha_\ell + \beta_\ell \frac{G-1}{H}\right)\frac{n_\ell}{n_0} H + aG \tag{4.138}$$

Notice that we have replaced the hazard functions in (4.44) and (4.45) by their Taylor approximation. Solving (4.137) and (4.138) for H and G we obtain

$$H = B + (B^2 + C)^{1/2} \tag{4.139}$$

$$G = \frac{a - \beta_\ell^*}{1 + \beta_r^* - \beta_\ell^*} - \frac{\alpha_1^* - \alpha_\ell^*}{1 + \beta_r^* - \beta_\ell^*} H \tag{4.140}$$

where

$$\alpha_r^* = \alpha_r n_r/n_0, \quad \alpha_\ell^* = \alpha_\ell n_\ell/n_0 \tag{4.141}$$

$$\beta_r^* = \beta_r n_r/n_0, \quad \beta_\ell^* = \beta_\ell n_\ell/n_0 \tag{4.142}$$

and

$$B = \frac{1}{2}\left[\alpha_\ell^* - (a - \beta_\ell^*)\frac{\alpha_r^* - \alpha_\ell^*}{1 + \beta_r^* - \beta_\ell^*}\right] \tag{4.143}$$

$$C = b + a^2 - \beta_\ell^* - \frac{(a - \beta_\ell^*)^2}{1 + \beta_r^* - \beta_\ell^*} \tag{4.144}$$

If censoring is symmetrical we obtain $\alpha_r = \alpha_\ell$, $\beta_r = -\beta_\ell$,

$$B = \frac{1}{2}\alpha_\ell^*, \quad C = b + a^2 - \frac{(a - \beta_\ell^*)^2}{1 + 2\beta_\ell^*} - \beta_\ell^* \tag{4.145}$$

and

$$G = \frac{a - \beta_\ell^*}{1 + 2\beta_\ell^*} \tag{4.146}$$

$$H = \frac{1}{2}\alpha_\ell^* + [\alpha_\ell^{*2}/4 + b + a^2 - (a - \beta_\ell^*)^2/(1 + 2\beta_\ell^*)]^{1/2} \tag{4.147}$$

We notice that Tiku's (1967) solution looks somewhat different to ours because he applied the Taylor approximation directly to the likelihood equations. But the estimates presented here are identical with Tiku's MMLEs.

The MML estimator was applied by Tiku (see Chapter 6 for references) to various test problems. In other papers Tiku (1980, 1983) proposed the MMLEs as robust estimators for location and scale parameters in parametric models other than normal. The basic idea is the same which led to the winsorized estimates (see, for instance, Dixon and Tukey, 1968), namely that nonnormality often comes from the tails of a distribution and if the extreme observations are censored then the difference between a normal and nonnormal sample diminishes.

Restricted Maximum Likelihood Estimators

Persson and Rootzen (1977) introduced particularly simple estimators for the mean and standard deviation of a normal distribution based on a Type I singly censored sample. These estimators are solutions to a restricted likelihood function (Schneider, 1984) and (Persson, 1984) extended Persson and Rootzen's results to doubly censored samples. By combining the method of maximum likelihood and the method of moments, estimators are derived which are explicit functions of the first and second sample moments and the percentages of censored values on the right and left. These estimators have high efficiency and the estimates turn out to be very close to the maximum likelihood estimates.

We consider a doubly Type I censored sample of size n where n_0 observations are available. Let $y_i = x_i - x_\ell$, $d = x_r - x_\ell$, $u_\ell = (x_\ell - \mu)/\sigma$ and $u_r = (x_r - \mu)/\sigma$, then the log likelihood function of the censored sample is up to a constant

$$\log L = n_\ell \log \Phi(u_\ell) + n_r \log[1 - \Phi(u_r)] - n_0 \log \sigma - \frac{1}{2} \sum_{i=n_\ell+1}^{n-n_r} (y_i/\sigma + u_\ell)^2 \tag{4.148}$$

where Φ is the standard normal distribution function. We assume $n_\ell \geqslant n_r$ and consider restricted maximum likelihood estimators by setting $\Phi(u_\ell) = n_\ell/n$. If $n_r > n_\ell$, we fix $\Phi(u_r) = 1 - n_r/n$. Hence

$$\frac{\partial \log L}{\partial \sigma} = n_r \frac{\phi(d/\sigma + u_\ell)d/\sigma^2}{[1 - \Phi(d/\sigma + u_\ell)]} \frac{n_0}{\sigma} + \sum_{i=n_\ell+1}^{n-n_r} (y_i^2 + \sigma u_\ell y_i)/\sigma^3 \tag{4.149}$$

where Φ is the standard normal density function. The solution to the likelihood equation has to be found by iteration. Instead of doing so we suggest setting $\Phi(u_r) = 1 - n_r/n$. Thus we are able to derive an explicit solution for σ^*:

$$\sigma^* = \frac{1}{2}C + \frac{1}{2}(C^2 + 4\overline{y}^2)^{1/2} \tag{4.150}$$

where $C = \overline{y}u_\ell + \phi(u_r)nd/n_0$, $\overline{y} = \Sigma y_i/n_0$ and $\overline{y}^2 = \Sigma y_i^2/n_0$. The method of moments is now applied to find an estimator for μ. Let $Z_r = \phi(u_r)/[\Phi(u_r) - \Phi(u_\ell)]$ and $Z_\ell = \phi(u_\ell)/[\Phi(u_r) - \Phi(u_\ell)]$; then the expected value of $\overline{X} = X_i/n_0$ is (see Chapter 2) $E(\overline{X}) = \mu + \sigma(Z_\ell - Z_r)$ and thus we obtain the estimator

$$\mu_{RML} = \overline{x} - \sigma^*(Z_\ell - Z_r) \tag{4.151}$$

Similarly, we can use the expected value of $S^2 = \Sigma X_i^2/n_0 - \bar{X}^2$, namely (asymptotically) $E(S^2) = \sigma^2[1 + u_\ell Z_\ell - u_r Z_r - (Z_\ell - Z_r)^2]$, to find an estimator for σ which is less biased than σ^* for small sample sizes.

$$\sigma_{RML} = \{s^2 - \sigma^{*2}[u_\ell Z_\ell - u_r Z_r - (Z_\ell - Z_r)^2]\} \tag{4.152}$$

Since the restrictions $\Phi(u_r) = 1 - n_r/n$ and $\Phi(u_\ell) = n_\ell/n$ are used, we obtain for Z_r and Z_ℓ the formulas

$$Z_r = \phi[\Phi^{-1}(1 - n_r/n)]n/n_0 \tag{4.153}$$

$$Z_\ell = \phi[\Phi^{-1}(n_\ell/n)]n/n_0 \tag{4.154}$$

which are to be used in (4.151) and (4.152). For singly right censored samples we obtain $Z_\ell = 0$ and let $p_0 = n_0/n$, $u_r = \Phi^{-1}(p_0)$, then

$$Z_r = \phi(u_r)/p_0 \tag{4.155}$$

Furthermore, let

$$g = u_r Z_r + Z_r^2 \tag{4.156}$$

$$C = -[x_{(n_0)} - \bar{x}]u_r \tag{4.157}$$

and

$$\sigma^* = \frac{1}{2}C + \frac{1}{2}[C^2 + 4s^2 + r(x_{(n_0)} - \bar{x})^2]^{1/2} \tag{4.158}$$

then

$$\mu_{RML} = x + \sigma^* Z_r \tag{4.159}$$

$$_{RML} = \{s^2 + \sigma^{*2}g\}^{1/2} \tag{4.160}$$

which are the Persson and Rootzen (1977) estimators. To illustrate the ease with which the RMLEs can be computed we prepared Table 4.8

TABLE 4.8 Factors for RMLE

n_r/n	0.9	0.8	0.7	0.6	0.5	0.4	0.3	0.2	0.1
Z_r	1.755	1.400	1.159	0.966	0.798	0.644	0.497	0.350	0.195
g	0.831	0.781	0.735	0.688	0.637	0.578	0.507	0.417	0.288
u_r	-1.282	-0.842	-0.524	-0.253	0.0	0.253	0.524	0.842	1.282

which gives the factors to be used in (4.159) and (4.160). But we notice that tables are not necessary since the estimates are in closed form. We notice that the restricted MLE can also be used for Type II censored samples.

Example 4.7 (One-sided censoring)

In Example 4.2 we computed the MLE for the inoculated mice data from Gupta (1952). The log-survival times are 1.613, 1.643, 1.663, 1.732, 1.740, 1.763, 1.778. The sample mean is 1.7048 and the sample variance is $s^2 = 0.003547$. Three of the 10 observations are censored at the right.

To compute the RMLEs for this data we first compute

$$C = -[x_{(n_0)} - x)u_r = -(1.778 - 1.7048)\ 0.524 = -0.03849$$

and from (4.158) we obtain $\sigma^* = 0.07721$.

Equations (4.159) and (4.160) now give the RMLEs

$$\mu_{RML} = 1.7048 + 0.7721\ 0.497 = 1.7432$$

and

$$\sigma_{RML} = (0.003547 + 0.07721^2\ 0.507) = 0.0811$$

where $Z_r = 0.497$ and $g = 0.507$ were taken from Table 4.8. For comparison we give the MLE and MML estimators.

$$\mu_{ML} = 1.7422, \quad \sigma_{ML} = 0.0792$$

$$\mu_{MML} = 1.7423, \quad \sigma_{MML} = 0.0794$$

Example 4.8 (Two-sided censoring)

In a time-mortality experiment (Cohen, 1957), the first observation is delayed until the collapse of a fixed time interval, with the result that n_ℓ sample specimens die before observation begins. The experiment is subsequently terminated at a predetermined time with n_r specimens remaining alive. Actual survival times are recorded for the n_0 specimens which die during the period of full observation. For a specific sample of this type, X, the log survival time in days, is assumed to be normally distributed. Hence the relevant parameters are $n_r = 5$, $n_\ell = 2$, $n = 47$, $\bar{x} = 1.62011$, $\overline{x^2} = 2.646499$, $x_\ell = 1.301030$, $x_r = 1.903090$. The estimates are

$$\hat{\mu}_{ML} = 1.6433, \qquad \hat{\sigma}_{ML} = 0.2028$$

$$\hat{\mu}_{RML} = 1.6426 \qquad \hat{\sigma}_{RML} = 0.2035$$

$$\mu_{MML} = 1.6433 \qquad \sigma_{MML} = 0.2029$$

Asymptotic variance of RMLE. Expanding $\sigma - \sigma^*(n_\ell/n, n_r/n, \bar{y}, \overline{y^2})$ in a Taylor series around $[E(n_\ell/n), E(n_r/n), E(\bar{y}), E(\overline{y^2})]$ demonstrates that σ^* is a consistent estimate of σ and $\sqrt{n}(\sigma^* - \sigma)$ is asymptotically normally distributed. The same holds for μ_{RML} and σ_{RML}. The asymptotic variances of the RML-estimators were calculated by a Taylor expansion.

$$\text{var}(\mu_{RML}) = \frac{\sigma^2}{n}\left[A_1^2 \text{var}(\bar{y})n + A_2^2 \text{var}(\overline{y^2})n + A_3^2 \text{var}\left(\frac{n_r}{n}\right)n \right.$$
$$+ A_4^2 \text{var}\left(\frac{n_\ell}{n}\right)n + 2(A_1 A_2 \text{cov}(\bar{y}, \overline{y^2})n$$
$$\left. + A_3 A_4 \text{cov}\left(\frac{n_r}{n}, \frac{n_\ell}{n}\right)n\right] + o(n^{-1}) \tag{4.161}$$

$$\operatorname{var}(\sigma_{RML}) = \frac{\sigma^2}{n}\left[B_1^2 \operatorname{var}(y)n + B_2^2 \operatorname{var}(y^2)n + B_3^2 \operatorname{var}\left(\frac{n_r}{n}\right)n\right.$$

$$+ B_4^2 \operatorname{var}\left(\frac{n_\ell}{n}\right)n + 2(B_1 B_2 \operatorname{cov}(\bar{y}, \bar{y}^2)n$$

$$\left. + B_3 B_4 \operatorname{cov}\left(\frac{n_r}{n}, \frac{n_\ell}{n}\right)n\right] + o(n^{-1}) \qquad (4.162)$$

Let

$$m_1 = m_1'\sigma = \sigma(Z_\ell - Z_r - u_\ell)$$

$$m_2 = m_2'\sigma^2 = \sigma^2[1 - Z_r(u_r - u_\ell) - u_\ell(Z_\ell - Z_r) + u_\ell^2]$$

$$m_3 = m_3'\sigma^3 = \sigma^3(u_r - u_\ell)^2 Z_r + 2\sigma^2 m_1 - u_\ell \sigma m_2$$

Then

$$\operatorname{var}(\bar{y}) = (m_2 - m_1^2)/E[n_0]$$

$$\operatorname{var}(\bar{y}^2) = (m_4 - m_2^2)/E[n_0]$$

$$\operatorname{var}\left(\frac{n_r}{n}\right) = \frac{n_r}{n}\left(1 - \frac{n_r}{n}\right)/n$$

$$\operatorname{var}\left(\frac{n_\ell}{n}\right) = \frac{n_\ell}{n}\left(1 - \frac{n_\ell}{n}\right)/n$$

$$\operatorname{cov}(\bar{y}, \bar{y}^2) = (m_3 - m_1 m_2)/E[n_0]$$

$$\operatorname{cov}\left(\frac{n_r}{n}, \frac{n_\ell}{n}\right) = -\frac{n_r n_\ell}{n^3}$$

Note that all other covariances are zero and thus they are omitted in the above equations.

$$A_1 = 1 - u_\ell(Z_\ell - Z_r)/(1 + m_2')$$

$$A_2 = -(Z_\ell Z_r)/[\sigma(1 + m_2')]$$

$$A_3 = -\left[\frac{(m_1' + Z_r Z_\ell \Delta)(Z_\ell - Z_r)}{1 + m_2'} + Z_\ell m_1'\right]\sigma/\phi(u_\ell)$$

$$A_4 = -\left[\frac{(Z_r^2 + Z_r u_r)\Delta(Z_\ell - Z_r)}{1 + m_2'} + Z_r(m_1' - \Delta)\right]\sigma/\phi(u_r)$$

$$B_1 = -[m_1' + u_\ell(m_2' - m_1'^2 - 1)/(1 + m_2')]$$

$$B_2 = [0.5 - (m_2' - m_1'^2 - 1)/(1 + m_2')]/\sigma$$

$$B_3 = -\sigma[(m_2' - m_1'^2 - 1)(m_1' + Z_r Z_\ell \Delta)/(1 + m_2')$$
$$+ 0.5 Z_\ell (m_2' - 2m_1'^2)]/\phi(u_\ell)$$

$$B_4 = -\sigma Z_r[(m_2' - m_1'^2 - 1)(Z_r + u_r)\Delta/(1 + m_2')$$
$$+ 0.5(m_2' - m_1'^2 - (Z_\ell - Z_r - u_r)^2]/\phi(u_r)$$

where $\Delta = u_r - u_\ell$.

Persson's Estimator

Persson (1983) gives estimators for doubly censored samples which are more complicated to compute than the RML estimator described above. But the asymptotic relative efficiency of Persson's estimators μ_{Per} and σ_{Per} are at least 99.47% and 99.27%, respectively. We shall give therefore the formulas derived by Persson (1983).

Let

$$u_\ell^* = \Phi^{-1}(n_\ell/n) \quad \text{and} \quad u_r^* = \Phi^{-1}(1 - n_r/n) \tag{4.163}$$

Furthermore we use Z_r and Z_ℓ as given by (4.153) and (4.154).

Persson considers Type I censoring. Define

$$a_r = 0.5\left\{\phi(u_\ell^*)(x_r - x_\ell)n/n_0 + u_r^*(\bar{x} - x_r) + \left[\left(\phi(u_\ell^*)(x_r - x_\ell)n/n_0 + u_r^*(\bar{x} - x_r)\right)^2 + (4/n_0)\sum_{i=n_\ell+1}^{n-n_r}(x_i - x_r)^2\right]^{1/2}\right\} \tag{4.164}$$

and

$$a_\ell = 0.5\left\{\phi(u_r^*)(x_r - x_\ell)n/n_0 + u^*(\bar{x} - x_\ell) + \left[\left(\phi(u_r^*)(x_r - x_\ell)n/n_0 + u_\ell^*(\bar{x} - x_\ell)\right)^2 + (4/n_0)\sum_{i=n_\ell+1}^{n-n_r}(x_i - x_\ell)^2\right]^{1/2}\right\} \tag{4.165}$$

Furthermore let

$$\sigma^* = [a_r\phi(u_r) + a_\ell\phi(u_\ell)]/[\phi(u_\ell) + \phi(u_r)] \tag{4.166}$$

then Persson's estimators are

$$\mu_{Per} = \bar{x} + Z_r a_r - Z_\ell a \tag{4.167}$$

and

$$\sigma^2_{Per} = s^2 - \sigma^{*2}[u_\ell^* Z_\ell - u_r^* Z_r - (Z_\ell - Z_r)^2] \tag{4.168}$$

4.5 BIAS CORRECTION

The maximum likelihood estimators are unfortunately biased. The bias tends to zero as n tends to infinity, but for small n the bias is significantly large to warrant consideration. The exact bias of σ_{ML} and μ_{ML} is unknown. For Type II singly censored samples Saw (1961b) found the leading term (i.e., the term of order $1/n$) in the bias of μ_{ML} and σ_{ML}.

Let $p_r^* - n_r/(n + 1)$; then the expectations $E[\ _{ML}]$ and $E[\mu_{ML}]$ can be expanded in terms of a power series in $1/(n + 1)$

$$E[\sigma_{ML}] = \sigma + \frac{\sigma}{n + 1} B(\sigma_{ML}, p_r^*) + O(n + 1)^{-2} \tag{4.169}$$

$$E[\mu_{ML}] = \mu + \frac{\sigma}{n + 1} B(\mu_{ML}, p_r^*) + O(n + 1)^{-2} \tag{4.170}$$

The values of $B(\mu_{ML}, p_r^*)$ and $B(\sigma_{ML}, p_r^*)$ are given in Tables 4.9 and 4.10.

Saw suggests constructing the corrected estimates

$$\mu_{ML}^c = \mu_{ML} - \frac{\sigma}{n + 1} B(\mu_{ML}, p_r^*) \tag{4.171}$$

$$\sigma_{ML}^c = \sigma_{ML} - \frac{\sigma}{n + 1} B(\sigma_{ML}, p_r^*) \tag{4.172}$$

in order to reduce the bias to order $1/(n + 1)^2$.

Saw's correction terms were only calculated for singly censored samples. Complicated calculations have to be done to obtain the leading term in the bias of μ_{ML} and σ_{ML} for doubly censored samples. Simulation studies show that the bias in μ_{ML} mainly depends on the degree of unsymmetry of censoring while the bias in σ_{ML} mainly depends on the overall percentage of censored observations. We therefore suggest using the following convenient approximation, which is based on a least square fit to the data in Table 4.9 and Table 4.10. Define

$$n_1 = n - 2 \times \min(n_\ell, n_r) \qquad n_0 = n - n_r - n_\ell \tag{4.173}$$

$$p_\mu = n_0/(n_1 + 1) \tag{4.174}$$

$$p_\sigma = n_0/(n + 1) \tag{4.175}$$

$$B(\hat{\mu}, p_\mu) = -\exp(2.692 - 5.439 p_\mu) \tag{4.176}$$

$$B(\sigma, p_\sigma) = -(0.312 + 0.859 p_\sigma)^{-2} \tag{4.177}$$

TABLE 4.9 Values of $B(\hat{\mu}, p_r^*)$, in $E(\hat{\mu}) = \mu + \sigma/(n + 1)B(\hat{\mu}, p_r^*) + O(n + 1)^{-2}$

p_r^*	$B(\hat{\mu} : p_r^*)$	p_r^*	$B(\hat{\mu} : p_r^*)$
0.20	-5.538	0.55	-0.750
0.25	-3.792	0.60	-0.583
0.30	-2.766	0.65	-0.450
0.35	-2.078	0.70	-0.342
0.40	-1.588	0.75	-0.254
0.45	-1.232	0.80	-0.181
0.50	-0.960	1.00	-0.000

Source: Saw (1961b).

TABLE 4.10 Values of $B(\hat{\sigma}, p_r^*)$, in $E(\hat{\sigma}) = \sigma + \sigma/(n + 1)B(\hat{\sigma}, p_r^*) + O(n + 1)^{-2}$

p_r^*	$B(\hat{\sigma} : p_r^*)$	p_r^*	$B(\hat{\sigma} : p_r^*)$
0.20	-4.777	0.55	-1.582
0.25	-3.751	0.60	-1.431
0.30	-3.087	0.65	-1.303
0.35	-2.614	0.70	-1.193
0.40	-2.258	0.75	-1.098
0.45	-1.983	0.80	-1.013
0.50	-1.762	1.00	-0.750

Source: Saw (1961b).

The estimators

$$\hat{\mu}^c = \hat{\mu} - \frac{\hat{\sigma}}{n+1} B(\hat{\mu}, p_\mu)\,\mathrm{sign}(n_r - n_\ell) \tag{4.178}$$

and

$$\hat{\sigma}^c = \hat{\sigma} - \frac{\hat{\sigma}}{n+1} B(\hat{\sigma}, p_\sigma) \tag{4.179}$$

are nearly unbiased, when the ML, MML or RML-estimators are used. Note that the formulas (4.178) and (4.179) can also be applied for doubly censored samples, where sign(0) = 0.

A much simpler technique can be applied to reduce the bias in σ_{ML}. As in the uncensored case we can use a degree of freedom adjustment to obtain an almost unbiased estimator for σ. That is,

$$\sigma_{ML}^a = \sigma_{ML}\sqrt{n_0/(n_0 - 1)} \tag{4.180}$$

where n_0 is the number of uncensored observations. This correction was also suggested by Tiku (1978) in order to adjust the MML estimator. We note that the bias of the RML estimator σ_{RML} can also be reduced in this way.

Although the bias corrections discussed above were derived for Type II censored samples, they will reduce the bias of Type I censored samples as well.

Simulation studies show (see Section 4.6) that formula (4.179) reduces the bias in σ_{ML} more than (4.180) for the cost of slightly higher variance. We also note that simulated bias corrections for σ_{ML} were given by Custer (1976).

Example 4.9

We consider Example 4.2. The maximum likelihood estimators for the parameters μ and σ of the lognormal distribution were found to be (see Example 4.2) $\mu_{ML} = 1.7422$ and $\sigma_{ML} = 0.0792$. The bias corrected MLEs are $\hat{\mu}^c = 1.7450$ and $\hat{\sigma}^c = 0.892$.

4.6 OPTIMAL SPACING

In situations where an extremely large sample is available, a linear estimate based on a few selected order statistics reduces the amount of computations. Some authors therefore deal with an asymptotic best linear unbiased estimator (ABLUE) for the mean and standard deviation, based on k selected order statistics. Ogawa (1951, 1973, 1974, 1977), Kulldorf (1961, 1963), and Chan (1970, 1973) have derived ABLU-estimators of the mean and the standard deviation from a complete sample. Cheng (1980) shows how those results can be extended to censored samples when σ is known. The results presented below are taken from Ogawa (1977).

A set of k real numbers satisfying the order relation

$$0 \equiv \lambda_0 < \lambda_1 < \lambda_2 < \cdots < \lambda_k < \lambda_{k+1} \equiv 1$$

is called a spacing of the k sample quantiles

$$x_{(n_1)} < x_{(n_2)} < \cdots < x_{(n_k)}$$

where

$$n_i = \begin{cases} n\lambda_i & \text{if } n\lambda_i \text{ is an integer} \\ [n\lambda_i + 1] & \text{otherwise} \end{cases}$$

Note that [y] is the integer part of y. Consider an uncensored sample and let

$$K_1 = \sum_{i=1}^{k+1} \frac{(f_i - f_{i-1})^2}{\lambda_i - \lambda_{i-1}} \tag{4.181}$$

$$K_2 = \sum_{i=1}^{k+1} \frac{(u_i f_i - u_{i-1} f_{i-1})^2}{\lambda_i - \lambda_{i-1}} \tag{4.182}$$

$$K_3 = \sum_{i=1}^{k+1} \frac{(f_i - f_{i-1})(u_i f_i - u_{i-1} f_{i-1})}{\lambda_i - \lambda_{i-1}} \tag{4.183}$$

$$\Delta = K_1 K_2 - K_3^2$$

$$u_i = \frac{x_i - \mu}{\sigma}, \quad f_i = \phi(u_i), \quad \lambda_i = \int_{-\infty}^{u_i} \phi(t)\, dt$$

Then the ABLU-estimators μ_{ABLU} and σ_{ABLU} for a given spacing are obtained by solving the system of equations (Ogawa, 1977)

$$\begin{aligned} K_1\hat{\mu} + K_3\hat{\sigma} &= X \\ K_3\hat{\mu} + K_2\hat{\sigma} &= Y \end{aligned} \tag{4.185}$$

where

$$X = \sum_{i=1}^{k+1} \frac{(f_i - f_{i-1})[f_i x_{(n_i)} - f_{i-1} x_{(n_{i-1})}]}{\lambda_i - \lambda_{i-1}} \tag{4.187}$$

Hence

$$\mu_{ABLU} = \frac{1}{\Delta}(K_2 X - K_3 Y) \equiv \sum_{i=1}^{k} c_i x_{(n_i)} \tag{4.188}$$

where

$$c_i = \frac{K_2}{\Delta} f_i \frac{f_i - f_{i-1}}{\lambda_i - \lambda_{i-1}} - \frac{f_{i+1} - f_i}{\lambda_{i+1} - \lambda_i} - \frac{K_3}{\Delta} f_i \left[\frac{u_i f_i - u_{i-1} f_{i-1}}{\lambda_i - \lambda_{i-1}} - \frac{u_{i+1} f_{i+1} - u_i f_i}{\lambda_{i+1} - \lambda_i} \right] \quad i = 1, 2, \ldots, k \tag{4.189}$$

and

$$\sigma_{ABLU} = \frac{1}{\Delta}(-K_3 X + K_1 Y) \equiv \sum_{i=1}^{k} d_i x_{(n_i)} \tag{4.190}$$

where

$$d_i = -\frac{K_3}{\Delta} f_i \left[\frac{f_i - f_{i-1}}{\lambda_i - \lambda_{i-1}} - \frac{f_{i+1} - f_i}{\lambda_{i+1} - \lambda_i} \right]$$

$$+ \frac{K_1}{\Delta} f_i \left[\frac{u_i f_i - u_{i-1} f_{i-1}}{\lambda_i - \lambda_{i-1}} - \frac{u_{i+1} f_{i+1} - u_i f_i}{\lambda_{i+1} - \lambda_i} \right] \quad i = 1, 2, \ldots, k \tag{4.191}$$

The asymptotic variance and covariance is then given as (Ogawa, 1977)

$$\text{var}(\mu_{ABLU}) = \frac{\sigma^2}{n} \frac{K_2}{\Delta} \tag{4.192}$$

$$\text{var}(\sigma_{ABLU}) = \frac{\sigma^2}{n} \frac{K_1}{\Delta} \tag{4.193}$$

$$\text{cov}(\mu_{ABLU}, \sigma_{ABLU}) = \frac{\sigma^2}{n} \frac{K_3}{\Delta}$$

It can be seen that the asymptotic variance and covariance depend on the spacing. When k is given but not the spacing itself the problem of choosing the best spacing arises in the sense that the asymptotic variance is minimized.

σ^2 known. We consider now the case that σ is known. Ogawa (1951) finds the asymptotically best linear unbiased estimate μ of the location parameter μ when the scale parameter σ is known.

$$\mu^* = \sum_{i=1}^{k} a_i x_{(n_i)} - \frac{K_3}{K_1} \sigma \tag{4.194}$$

where

$$a_i = \frac{f_i}{K_1} \left[\frac{f_i - f_{i-1}}{\lambda_i - \lambda_{i-1}} - \frac{f_{i+1} - f_i}{\lambda_{i+1} - \lambda_i} \right] \tag{4.195}$$

$$K_1 = \sum_{i=1}^{k+1} \frac{(f_i - f_{i-1})^2}{\lambda_i - \lambda_{i-1}} \tag{4.196}$$

$$K_3 = \sum_{i=1}^{k+1} \frac{(f_i - f_{i-1})(f_i u_i - f_{i-1} u_{i-1})}{\lambda_i - \lambda_{i-1}} \tag{4.197}$$

$$u_i = \frac{x_i - \mu}{\sigma}, \quad f_i = \phi(u_i), \quad \lambda_i = \int_{-\infty}^{u_i} \phi(t)\, dt$$

$$f_0 \equiv f_{k+1} \equiv f_0 u_0 \equiv f_{k+1} u_{k+1} \equiv 0, \quad \lambda_0 \equiv 0, \quad \lambda_{k+1} \equiv 1$$

The spacing which minimizes $\text{var}(\mu^*) = \sigma^2/(nK_1)$, equivalently maximizing K_1, is called the optimum spacing

Complete sample. In order to find μ_{ABLU} and the optimum spacing $\{\lambda_i^c\}$ we have to maximize K_1 with respect to $0 < \lambda_1 < \lambda_2 < \cdots < \lambda_k < 1$. The following theorem is due to Ogawa (1951, (1974) and Higuchi (1956).

Theorem. There is a unique spacing $\{\lambda_i^c\}$ for the system of equations

$$M(\lambda_{i-1}, \lambda_i, \lambda_{i+1}) = \frac{f_{i+1} - f_i}{\lambda_{i+1} - \lambda_i} + \frac{f_i - f_{i-1}}{\lambda_i - \lambda_{i-1}} - 2\frac{df_i}{d\lambda_i} = 0$$

$$i = 1, 2, \ldots, k$$

which actually maximizes K_1.

The values of a_i, and $\{\lambda_i^c\}$ were calculated by Ogawa (1974).

Censored sample. Assume that a sample is doubly censored, where the largest $100p_r$% and the smallest $100p_\ell$% of the order statistics are missing. It is immediately evident that the optimal spacing of a complete sample is not affected by censoring if $p_\ell \leq \lambda_1^c$ and $1 - p_r \leq \lambda_k^c$. Cheng (1980) has shown that for cases where $p_\ell > \lambda_1^c$

or $1 - p_r < \lambda_k^c$ a unique spacing which minimizes $\text{var}(\mu_{ABLU})$ is obtained by setting $\lambda_1^c = p_\ell$ and $\lambda_k^c = 1 - p_r$, respectively, and solving the remaining system of (k - 2) equations

$$M(\lambda_{i-1}, \lambda_i, \lambda_{i+1}) = 0 \quad i = 2, 3, 4, \ldots, k - 1$$

For singly censored samples with $p_\ell > \lambda_1^c$ or $1 - p_r > \lambda_k^c$, one sets $p_\ell = \lambda_1^c$ or $1 - p_r = \lambda_k^c$, respectively, and solves the remaining k - 1 equations to obtain a unique spacing which minimizes $\text{var}(\mu_{ABLU})$.

The ABLU estimator has been compared with the MLE. The asymptotic efficiency of μ_{ABLU} relative to μ_{ML} is at least 94% if 5 or more order statistics in the censored sample are used. But unfortunately small sample efficiencies have not been investigated yet. Cheng (1980) prepared a table for the coefficients of the optimum spacing.

4.7 COMPARISON OF ESTIMATORS

In this section we shall compare several of the estimators discussed in the preceding sections. The term "asymptotic relative efficiency" (ARE) will be used with respect to the ratio of the asymptotic variance of an estimator and the Cramer-Rao bound. In the finite sample case we shall use the term "efficiency of an estimator" with respect to the ratio of the mean square error and the mean square error of the MLE. If we compare only linear estimators we shall take the variance of an estimator over the variance of the BLU estimator as a measure of efficiency.

Asymptotic Relative Efficiency

Most of the estimators discussed in the preceding sections are asymptotically efficient. The asymptotic variance of the MLE can be obtained from Table 3 of Appendix A. Tiku's MML and Chernoff's estimator are asymptotically efficient as shown by Tiku (1967) and Chernoff et al. (1967), respectively. The asymptotic efficiency of the BLU estimator was demonstrated by Plackett (1958). The ARE of the RML estimators $\hat{\mu}$ and $\hat{\sigma}$ are greater than 99.6% and 98%, respectively, when the sample is singly censored (Persson and Rootzen, 1977). For doubly censored samples the ARE of the RML estimator $\hat{\mu}$ drops to 94.93% for $u_\ell = -0.51$ and $u_r = 0.51$, while the ARE of $\hat{\sigma}$ takes its minimum (93.22%) for $u_\ell = -1.25$ and $u_r = 1.25$. The ARE returns to 100% as the censoring tends to totality (Schneider, 1984). Persson's (1983b) estimator $\hat{\mu}$ and $\hat{\sigma}$ have AREs of 99.45% and 99.6%, respectively.

Small Sample Efficiency

The small sample efficiency of estimators for Type I and Type II censored samples has been studied by several authors. Dixon (1960) gave efficiencies of his estimators with respect to the BLU estimators. Harter and Moore (1966) studied the performance of the maximum likelihood estimators in the case of Type II censoring. They compared the mean square error of the MLE with the variance of the BLU estimator. A similar study had been done by Isida and Tagami (1959). Tiku (1970) gave simulation results for the MML estimator. Persson and Rootzen (1977) and Schneider (1984) performed simulation studies of the RML estimator and Persson (1983b) presented simulation results for his estimator. Schneider and Weissfeld (1985) studied the bias corrected MLE and RML estimators. The Chernoff estimator was compared with the MLE in a simulation study performed by Schneider (1985). We shall summarize the results of the various studies without giving extensive tables of simulation results. A comprehensive simulation study which includes all the estimators mentioned above is given by Schneider (1985).

The loss of efficiency through censoring in the estimation of μ depends highly on the symmetry of the censoring. Table 4.11 gives the efficiency of the best linear unbiased estimator in relation to the uncensored case. If a sample of size n = 10 is symmetrically censored where $n_r = n_\ell = 4$, the efficiency of the BLU-estimator is 72% compared with an uncensored sample. If the

TABLE 4.11 Efficiency of BLU Estimator μ_{BLU} in Comparison to an Uncensored Sample Estimator, n = 10

$n_\ell \backslash n_r$	0	1	2	3	4	5	6	7	8
0	1	0.98	0.93	0.86	0.75	0.60	0.42	0.24	0.09
1		0.96	0.92	0.86	0.75	0.58	0.37	0.16	
2			0.90	0.85	0.75	0.57	0.29		
3				0.82	0.74	0.54			
4					0.72				

TABLE 4.12 Efficiency of the BLU Estimator σ_{BLU} in Comparison to an Uncensored Sample Estimator

$n\backslash n_0$	2	3	4	5	6	7	8	9	10
2	100								
3	45	100							
4	25	60	100						
5	20	40	70	100					
6	15	30	50	75	100				
7	11	25	40	60	75	100			
8	9	20	35	45	60	80	100		
9	8	17	30	40	50	65	80	100	
10	7	15	25	35	45	55	70	85	100

Source: From Sarhan and Greenberg (1962).

8 largest (or smallest) observations are censored, the efficiency drops to 9%. The middle values contain the highest information for estimating μ. If the median is censored $n_r/n \geqslant 50\%$, then the efficiency drops under 60%. For estimating μ, single censoring above 50% should be avoided.

The loss of efficiency through censoring in estimating σ depends mainly on the percentage of censored observations as was pointed out by Sarhan and Greenberg (1962). For n = 10 the censoring of 4 observations results in a loss of efficiency of about 55%, regardless of the censoring scheme. Censoring has a much stronger effect on estimating σ than on estimating μ. If two observations out of ten are censored, the efficiency compared with an uncensored sample drops immediately to 70%. Table 4.12 gives the approximate efficiency of the BLU-estimator compared with an uncensored sample. Note that the loss of efficiency mainly depends on $n_0 = n - n_r - n_\ell$. The table therefore reports the approximate efficiencies dependent on n_0. Within the class of linear unbiased estimators the simplified estimators of Gupta (1952) and Dixon (1960) have high efficiency compared with the BLU estimator (see Gupta, 1952, and Dixon, 1960). The efficiency of Dixon's estimator of the mean does not fall below 95.6%. For symmetrical or singly censored samples Dixon's estimators are more efficient than Gupta's estimators (see Tables 4.13 and 4.14).

The estimator of the mean given by Chernoff et al. (1967) has high efficiency for symmetrically censored samples. For unsymmetrical censoring the latter estimator becomes rather biased. Chernoff's estimator of the standard deviation is heavily biased.

The performance of the ML, MML, and RML estimators in small samples is basically the same. The mean square error of the MLE, as pointed out by Harter and Moore (1966), is usually smaller than the variance of the BLU estimator for sample sizes larger than 10. The same holds for the MML and RML estimators in comparison to the BLU estimator. For moderate censoring the bias-corrected estimators (4.178) and (4.179) are almost unbiased and have a mean square error which is slightly smaller than the variance of the BLU estimator. Bias correction methods necessarily increase the variance of the estimator and it seems that the overall variation (MSE) is increased. Hence, if the MSE is the criterion for choosing an estimator the original MLE (or RMLE or MMLE) is the best estimator.

TABLE 4.13 Small Sample Efficiency of Estimators of Mean (n = 10)

						BLU	Gupta	Dixon	
		MLE		RMLE				w	a
n_ℓ	n_r	Bias	MSE	Bias	MSE	Var	Eff	Eff	Eff
0	1	-0.013	0.105	0.009	0.105	0.103	0.99	0.98	1.0
0	2	-0.026	0.110	0.006	0.110	0.107	0.98	0.97	1.0
0	3	-0.037	0.116	0.003	0.117	0.117	0.96	0.96	0.99
0	4	-0.059	0.133	0.009	0.136	0.134	0.94	0.97	0.98
0	5	-0.097	0.161	0.005	0.163	0.166	0.91	0.97	-
0	6	-0.168	0.224	-0.014	0.236	0.237	0.90	0.96	-
0	7	-0.289	0.357	-0.063	0.385	0.417	0.92	-	-
0	8	-0.559	0.739	-0.291	0.803	1.127	1.00	-	-

Note: Eff = Var/Var(BLU); the RMLE is bias corrected; the results are based on 10,000 samples.

TABLE 4.14 Efficiency of Estimators of the Standard Deviation (n = 10)

		Corrected RMLE						
		(1)		(2)		BLU	Gupta	Dixon
n_ℓ	n_r	Bias	MSE	Bias	MSE	Var	Eff	Eff
0	1	0.038	0.065	0.002	0.069	0.068	0.97	0.97
0	2	0.043	0.078	0.004	0.083	0.081	0.94	0.98
0	3	0.048	0.090	0.005	0.099	0.099	0.92	0.99
0	4	0.065	0.115	0.003	0.126	0.124	0.91	0.98
0	5	0.083	0.138	0.017	0.154	0.161	0.90	0.98
0	6	0.117	0.187	0.039	0.203	0.225	0.91	0.98
0	7	0.167	0.272	0.095	0.293	0.354	0.93	
0	8	0.272	0.469	0.253	0.483	0.749	1.00	
1	1	0.143	0.077	0.012	0.087	0.082	0.97	
2	2	0.075	0.117	0.003	0.130	0.129	0.96	
3	3	0.117	0.203	0.037	0.231	0.244	0.96	
4	4	0.292	0.511	0.280	0.520	0.846	1.00	

Note: Eff = Var/Var(BLU); (1) corrected with (4.180); (2) corrected with (4.179); the results are based on 10,000 samples.

If an unbiased estimator is required then the bias-corrected MLE, MMLE or RMLE can be used even for small samples. The reduction of the bias using Eqs. (4.178) and (4.179) is substantial, as Tables 4.13 and 4.14 show. If interval estimation or tests are considered, bias corrected estimators should be used (see Chapter 6) to obtain nearly unbiased confidence intervals and tests.

We finally mention that the knowledge of one parameter, μ or σ, reduces the variance and bias of the estimators considerably (see Tiku, 1967, and Harter, 1969).

We notice that the conclusions regarding the MLE, RMLE, and MMLE are basically the same and hold for Type I and Type II censoring.

REFERENCES

Ali, M. M. and Chan, L. K. (1964). On Gupta's estimates of the parameter of the normal distribution, Biometrika 51, 498-501.

Bliss, C. I. (1937). The calculation of the time-mortality curve, Annals of Applied Biology 24, 815-852.

Bose, R. C. and Gupta, A. K. (1959). Moments of order statistics from a normal population, Biometrika 46, 433-439.

Breakwell, J. V. (1953). On estimating both mean and standard deviation of a normal population from the lowest r out of n observations (Abstract), Annals of Mathematical Statistics 24, 683.

Brooks, R. J. (1982). On the loss of information through censoring, Biometrika 69, 137-144.

Chan, L. K. (1970). On the best linear estimate of the mean or standard deviation of the normal distribution based on order statistics, Journal of Statistical Research 4, 71-92.

Chan, L. K. and Chan, N. N. (1973). On the optimum best linear estimates of the parameters of the normal distribution based on selected order statistics, Scandinavian Actuarial Journal 56, 120-128.

Cheng, W. (1980). On the ABLUE of the normal mean from a censored sample, Journal of Statistical Planning and Inference 4, 259-265.

Chernoff, H., Gastwirth, J. L., and Johns, M. V. (1967). Asymptotic distribution of linear combinations of functions of order statistics with application to estimation, Annals of Mathematical Statistics 38, 52-72.

Cohen, A. C. (1950). Estimating the mean and variance of normal populations from singly truncated and doubly truncated samples, Annals of Mathematical Statistics 21, 557-569.

Cohen, A. C. (1955). Censored samples from truncated normal distributions, Biometrika 42, 516-519.

Cohen, A. C. (1957). On the solution of estimating equations for truncated and censored samples from normal populations, Biometrika 44, 225-236.

Cohen, A. C. (1959). Simplified estimators for the normal distribution when samples are singly censored or truncated, Technometrics 1, 217-237.

Cohen, A. C., Jr. (1961). Tables for maximum-likelihood estimates; singly truncated and singly censored samples, Technometrics 3, 535-541.

Cooley, C. G. and Cohen, A. C. (1970). Tables of maximum likelihood estimating functions for singly truncated and singly censored samples from the normal distribution, Technical Report #54, Department of Statistics, University of Georgia, Athens.

Custer, S. W. (1976). Correction for bias in maximum likelihood estimators of σ in a right-censored normal distribution, Communications in Statistics, Simulation and Computation B5, 15-22.

David, F. N. and Johnson, N. L. (1954). Statistical treatment of censored data, Biometrika 41, 228-240.

David, H. A. (1968). Gini's mean difference rediscovered, Biometrika 55, 573-575.

Dixon, W. J. (1957). Estimates of the mean and standard deviation of a normal population, Annals of Mathematical Statistics 28, 806-809.

Dixon, W. J. (1960). Simplified estimation from censored normal samples, Annals of Mathematical Statistics 31, 385-391.

Dixon, W. J. and Tukey, J. W. (1968). Approximate behavior of the distribution of winsorized t (Trimming/Winsorization 2), Technometrics 10, 83-98.

Downton, F. (1966). Linear estimates with polynomial coefficients, Biometrika 53, 129-141.

Govindarajulu, F. and Rao, J. N. K. (1968). Asymptotically best estimator of scale based on two symmetric quantiles, Sankhya, Ser. A 30, 335-336.

Grundy, P. M. (1952). The fitting of grouped truncated and grouped censored normal distributions, Biometrika 39, 252-259.

Gupta, A. K. (1952). Estimation of the mean and standard deviation of a normal population from a censored sample, Biometrika 39, 260-273.

Hald, A. (1949). Maximum likelihood estimation of the parameters of a normal distribution which is truncated at a known point, Scandinavian Actuarial Journal 32, 119-134.

Halperin, M. (1952). Maximum likelihood estimation in truncated samples, Annals of Mathematical Statistics 23, 226-238.

Harter, H. L. (1961). Expected values of normal order statistics, Biometrika 48, 151-165.

Harter, H. L. (1970). Order statistics and their use in testing and estimation, Vol. 2, Washington, U.S. Government Printing Office.

Harter, H. L. and Moore, A. H. (1966). Iterative maximum likelihood estimation of the parameters of normal populations from singly and doubly censored samples, Biometrika 53, 205-213.

Healy, M. J. R. (1978). A mean difference estimator of standard deviation in symmetrically censored samples, Biometrika 65, 643-646.

Higuchi, I. (1956). On the solution of certain simultaneous equations in the theory of systematic statistics, Annals of the Institute of Statistical Mathematics 5, 77-90.

Isida, M. and Tagami, S. (1959). The bias and precision in the maximum likelihood estimation of the parameters of normal populations from singly truncated sample, Reports of Statistical Application Research Union of Japanese Scientists and Engineers 6, 105-110.

Jung, J. (1955). On linear estimates defined by a continuous weight function, Arikv for Mathematik, Band 3, 199-209.

Kulldorf, G. (1961). Estimation from grouped and partially grouped samples, Almquist and Wiksell, Stockholm.

Lloyd, E. H. (1952). Least squares estimation of location and scale parameters using order statistics, Biometrika 39, 88-95.

Mehrotra, K. G. and Nanda, P. (1974). Unbiased estimation of parameters by order statistics in the case of censored samples, Biometrika 61, 601-606.

Nelson, F. D. (1982). Applied Life Data Analysis, New York, Wiley.

Ogawa, J. (1951). Contributions to the theory of systematic statistics, I, Osaka Mathematical Journal 3, 175-213.

Ogawa, J. (1973). A further note on the optimal spacing of the systematic statistics: Estimation of the location parameter of a normal distribution, Mimeographed report, University of Calgary, Canada.

Ogawa, J. (1974). Asymptotic theory of estimation of the location and scale parameters based on selected sample quantiles (revised version), Mimeographed report, University of Calgary, Canada.

Ogawa, J. (1977a). A note on the optimal spacing of systematic statistics — normal distribution, in Essays in Probability and Statistics (in honor of J. Ogawa), Shinko Tsusho Co., Tokyo, 467-474.

Ogawa, J. (1977b). Optimal spacing for the simultaneous estimation of the location and scale parameters of a normal distribution based on selected two sample quantiles, Journal of Statistical Planning and Inference 1, 61-72.

Owen, D. B. (1966). On the distribution of the range and mean range for samples from a normal distribution, Biometrika 53, 245.

Persson, T. (1983). Simple and highly efficient estimators for a Type I doubly censored normal sample, Technical Report No. 5, University of Lund, Sweden.

Persson, T. (1984). Parameter estimation in censored samples and processes, Dissertation, University of Lund, Sweden.

Persson, T. and Rootzen, H. (1977). Simple and highly efficient estimators for a Type I censored normal sample, Biometrika 64, 123-128.

Plackett, R. L. (1958). Linear estimation from censored data, Annals of Mathematical Statistics 29, 131-142.

Prescott, P. (1970). Estimation of the standard deviation of a normal population from doubly censored samples using normal scores, Biometrika 57, 409-419.

Prescott, P. (1979). A mean difference estimator of standard deviation in asymmetrically censored normal samples, Biometrika 66, 684-686.

Sarhan, A. E. and Greenberg, B. G. (1956). Estimation of location and scale parameters by order statistics from singly and doubly censored samples, Annals of Mathematical Statistics 27, 427-457.

Sarhan, A. E. and Greenberg, B. G. (1958). Estimation of location and scale parameters by order statistics from singly and doubly censored samples, Part II. Tables for the normal distribution for samples of size 11 N 15, Annals of Mathematical Statistics 29, 79-105.

Sarhan, A. E. and Greenberg, B. G. (eds.) (1962). Contribution to Order Statistics, New York, Wiley.

Sarndal, C. E. (1962). Information from Censored Samples, Almquist and Wiksell, Stockholm.

Saw, J. G. (1958). Moments of sample moments of censored samples from a normal population, Biometrika 45, 211-221.

Saw, J. G. (1959). Estimation of the normal population parameters given a singly censored sample, Biometrika 46, 150-155.

Saw, J. G. (1961a). Estimation of the normal population parameters given a Type I censored sample, Biometrika 48, 367-377.

Saw, J. G. (1961b). The bias of the maximum likelihood estimates of location and scale parameters given a Type II censored normal sample, Biometrika 48, 448-451.

Saw, J. G. (1962). Efficient moment estimators when the variables are dependent with special reference to Type II censored normal data, Biometrika 49, 155-161.

Schneider, H. (1984). Simple and highly efficient estimators for censored normal samples, Biometrika 71, 412-414.

Schneider, H. (1985). A comparison of estimators based on censored samples from normal distribution, Technical Report, Department of Mathematics, University of Arkansas.

Schneider, H. and Weisfeld, L. (1986). Inference based on Type II censored samples, to appear in Biometrics.

Schmee, J. and Nelson, W. (1977). Estimates and approximate confidence limits for (log) normal life distributions from singly censored samples by maximum likelihood, General Electric Company Corporate Research and Development TIS Report #76CRD250.

Selvin, S. (1976). A graphical estimate of the population mean from censored normal data, Applied Statistics 25, 8-11.

Stevens, W. L. (1937). The truncated normal distribution appendix to "The calculation of the time-mortality curve" by C. I. Bliss, Annals of Applied Biology 24, 815-852.

Swamy, P. S. (1960). Estimating the mean and variance of a normal distribution from singly and doubly truncated samples of grouped observations, Bulletin of the Calcutta Statistical Association 9, 145-156.

Swamy, P. S. (1961). Estimating the parameters of Pearson Type III populations from singly and doubly truncated or censored samples of grouped observations, Journal of Indian Society of Agricultural Statistics 13, 147-158.

Swamy, P. S. (1962a). On the joint efficiency of the parameters of normal populations based on singly and doubly truncated samples, Journal of the American Statistical Association 57, 46-53.

Swamy, P. S. (1962b). On the amount of information supplied by censored samples of grouped observations in the estimation of statistical parameters, Biometrika 49, 245-249.

Swamy, P. S. (1963). On the amount of information supplied by truncated samples of grouped observations in the estimation of the parameters of normal populations, Biometrika 50, 207-213.

Swan, A. V. (1969a). Computing maximum likelihood estimates for parameters of the normal distribution from grouped and censored data, Applied Statistics 18, 65-69.

Swan, A. V. (1969b). Algorithm AS16. Maximum likelihood estimation from grouped and censored normal data, Applied Statistics 18, 110-114.

Swan, A. V. (1977). A remark on algorithm AS16. Maximum likelihood estimation from grouped and censored normal data, Applied Statistics 26, 122.

Teichroew, D. (1956). Tables of expected values of order statistics and products of order statistics for samples of size twenty and less from the normal distribution, Annals of Mathematical Statistics 27, 410-425.

Tiku, M. L. (1967). Estimating the mean and standard deviation from a censored normal sample, Biometrika 54, 155-165.

Tiku, M. L. (1968). Estimating the parameters of log-normal distribution from censored samples, Journal of the American Statistical Association 63, 134-140.

Tiku, M. L. (1970). Monte Carlo study of some simple estimators in censored normal samples, Biometrika 57, 207-210.

Tiku, M. L. (1971). Estimating the means and standard deviation from two censored normal samples, Biometrika 58, 241-243.

Tiku, M. L. (1980). Robustness of MML estimators based on censored samples and robust test statistics, Journal of Statistical Planning and Inference 4, 123-143.

Tiku, M. L. (1983). Exact efficiencies of some robust estimators in sample survey, Communication in Statistics — Theory and Methods 12, 2043-2051.

Van Zwet, W. R. (1966). Bias in estimation from Type I censored samples, Statistica Neerlandica 20, 143-148.

Walsh, J. E. (1956). Estimating population mean, variance, and percentage points from truncated, Scandinavian Actuarial Journal 39, 47-58.

Wolynetz, M. S. (1979). Maximum likelihood estimation from confined and censored normal data, Algorithm AS138, Applied Statistics 28, 185-195.

CHAPTER 5

Linear Regression with Censored Data

We consider the regression model defined by

$$Y_j = \underline{\beta}'\underline{x}_j + \epsilon_j \qquad j = 1, \ldots, n \tag{5.1}$$

where β is a k + 1 component column vector $\beta' = (\beta_0, \beta_1, \ldots, \beta_k)$ of unknown constants, $\underline{x}_j' = (1, x_{j1}, x_{j2}, \ldots, x_{jk})'$, $j = 1, \ldots, n$ are k + 1-component vectors of known constants and ϵ_j is a normally distributed random variable with expected value $E[\epsilon_j] = 0$, variance $\text{var}[\epsilon_j] = \sigma^2$ and $E[\epsilon_j \epsilon_{j'}] = 0$ for $j \neq j'$. We shall consider censoring of the dependent variable Y_j.

The linear model with censored data was considered by many authors from different fields such as economics, biometrics, and life testing. Some of the key papers are mentioned here. Further references can be found in the various sections below.

Tobin (1958) discussed an application of the censored linear model to an economical problem and Glasser (1965) considered a biometrical application. Amemiya (1973) derived the maximum likelihood estimator for the censored linear model and showed that the MLE is asymptotic normally distributed. At the same time Nelson and Hahn (1972, 1973) gave in two papers the theoretical derivation of linear estimation for the censored linear model and some applications in life testing. Dempster et al. (1977) described the expectation maximization algorithm to obtain maximum likelihood estimates. While Dempster et al. dealt with a general parametric model, Aitkin (1981) applied the EM algorithm to obtain estimates in the censored linear model with normal errors. Wolynetz (1979) gave a computer program for computing the MLE via the EM algorithm. Simplified estimators were suggested by Schmee and Hahn (1979) and by Tiku (1978). Schmee and Hahn (1981) also gave a computer program for computing their simplified estimators.

Example 5.1

An example of a censored linear model is given by Crawford (1970) and also discussed by Schmee and Hahn (1979). To test the electrical insulation of motorettes, accelerated life tests are performed. Forty motorettes were tested, in groups of ten at four different temperatures. Testing was terminated at different times at each temperature (Type I censoring). The model to analyze the data, often referred to as the Arrhenius model, assumes that: (i) for any temperature, the distribution of time to failure is lognormal; (ii) the standard deviation σ of the lognormal time to failure distribution is constant; and (iii) the mean of the logarithm of the time to failure is a linear function of the reciprocal $x = 1/T$ of the absolute temperature T. Hence the linear model $Y = \log t = \beta_0 + \beta_1 x + \epsilon$ applies, where t is the failure time. The dependent variable Y is Type I censored. Type II censoring might also be used in life testing but is less often applied. We shall analyze Crawford's example in more detail at the end of this chapter.

Example 5.2

The Stanford Heart Transplant data were considered by several authors. Miller and Halpern (1982), for instance, give survival times of potential heart transplant recipients from their date of acceptance into the Stanford Heart Transplant Program. One problem of considerable interest is the evaluation of the effect of age at transplantation on survival time. The data presented by Miller and Halpern (1982) reported the situation in February 1980. Some of the patients died; others were still alive; thus, their life time is censored. A patient has to wait until a suitable donor is located and transplantation takes place, at which time he joins the treatment group. Hence, a patient's joining the group occurs randomly; subsequently, the censoring time can also be considered random. In literature (see Miller and Halpern 1982) several nonparametric methods have been used to estimate the slope and intercept of the model log life $= \beta_0 + \beta_1$ Age $+ \epsilon$. At the end of this chapter we shall compare the nonparametric method of Buckley and James (1979) with the maximum likelihood estimator based on Type I censoring.

Example 5.3

Tobin (1958) considers as an example the Engle curve relationship showing how household expenditures on various categories of goods vary with household income. For many categories (luxuries) there are no expenditures for low income levels. A single straight line cannot, therefore, represent the Engle curve for both low and high incomes. If individual households were identical except for income level, the Engle curve would be a broken line as in Figure 5.1.

In this example the dependent variable of the regression has a lower limit, namely zero. Consider a sample of size n, (x_j, y_j), j = 1 ..., n; then y_j might take on the value zero for a number of observations and take a wider range of values for the remaining observations. The model can be described by

$$y_j = \begin{cases} \underline{\beta}'\underline{x}_j + \epsilon_j & \text{if RHS} > 0 \\ 0 & \text{otherwise} \end{cases} \tag{5.2}$$

We notice that, strictly speaking, this model is neither truncated nor censored. The term truncation was used in Chapter 3 for samples

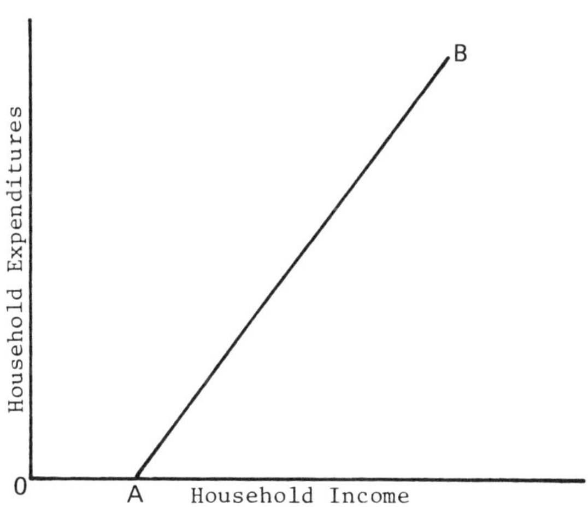

FIGURE 5.1 Expenditure as a function of income.

from truncated normal distribution, i.e., some value ranges of the population are excluded from observation. In Tobin's (1958) model negative y are not excluded but rather set to zero. Hence, the distribution of Y will have a positive probability at y = 0. The probability distribution of Y is therefore

$$f(y) = \begin{cases} \phi[(y_j - \underline{\beta}'\underline{x}_j)/\sigma] & y_j > 0 \\ \Phi(-\underline{\beta}\underline{x}_j/\sigma) & y_j = 0 \end{cases} \qquad j = 1, \ldots, n \tag{5.3}$$

which is not truncated normal. In order to apply a censored linear model we have to imagine that Y can take on negative values, but the observations are censored at zero. This was done by Tobin (1958), Amemiya (1973), Olsen (1980), and Nelson (1981). The problem with this approach is that the model is only linear in the interval (A, ∞). Thus we have to assume that we estimate within this interval, i.e., $\underline{\beta}'\underline{x}_j > 0$. But usually we do not know for what $\underline{x}_j$ the mean $\underline{\beta}'\underline{x}_j$ will become negative. We might ignore this and apply a censored linear model as an approximation, but we should notice that observations where $\underline{\beta}'\underline{x}_j < 0$ do not offer any information concerning β.

Example 5.4

Scott and Jones (1985) discuss a problem from agronomy. The purposes of their investigation were to quantify the effect of a wide range of suboptimal temperatures (i.e., 6-20°C) on seed germination and to compare responses among several high altitude ecotypes of wild Lycopersicon species and selected cultivated accessions (cultivated tomato). As low temperature limit for germination is approached, response times increase rapidly. As a result, some viable seeds commonly fail to germinate before an experiment is terminated. Those viable, nonresponding seeds are Type I censored observations.

First we shall present linear unbiased estimators for Type II censoring in Section 5.1. The maximum likelihood estimators are described in Section 5.2, and Section 5.3 deals with simplified estimators. Section 5.4 compares some estimators and discusses the effect of heteroscedasticity on the estimation of $\underline{\beta}$.

5.1 BEST LINEAR UNBIASED ESTIMATOR FOR TYPE II CENSORED SAMPLES

We consider the linear model defined by (5.1), i.e.,

$$Y_j = \underline{\beta}'\underline{x}_j + \epsilon_j \quad j = 1, \ldots, n \tag{5.4}$$

The dependent variable Y is normally distributed with mean μ and variance σ^2 where both these values are unknown. Most of the following results appear in Nelson and Hahn (1973).

Suppose that independent random samples are taken at each of M different conditions and that for the sample at the jth condition (j = 1, 2, ..., M) the values of the independent variables are $x_{j1}, \ldots, x_{jk}$. Then the (unknown) value of the location parameter is

$$\beta_0 + \beta_1 x_{j1} + \cdots + \beta_k x_{jk} = \mu_j \tag{5.5}$$

Suppose the size of the sample at the jth condition is n_j, i.e., $(y_{j1}, \ldots, y_{jn_j})$ and the sample is Type II censored. We consider censoring from above (on the right), where n_{rj} observations of the sample at the jth condition are censored. Thus the $n_{0j} = n_j - n_{rj}$ smallest observations are available and the observed order statistics of this sample are

$$y_{j(1)} \leqslant y_{j(2)} \leqslant \cdots \leqslant y_{j(n_{oj})}$$

Note that censoring from below (left censoring) can be treated as censoring from above by changing the sign of the observations. More general censoring can be treated as well. The expectation and variance of the ℓth order statistic $Y_{j(\ell)}$ at the jth condition is

$$E[Y_{j(\ell)}] = \mu_j + z_{\ell,n_j}\sigma \tag{5.6}$$

and

$$\text{var}[Y_{j(\ell)}] = \sigma^2 v_{n_j\ell} \tag{5.7}$$

where z_{ℓ, n_j} and $v_{n_j \ell}$ are the expectation and variance of the ℓth order statistic from a standard normal distribution. These constants are independent of μ and σ but they depend on the sample size n_j. The covariance of the order statistics $Y_{j(\ell)}$ and $Y_{j(\ell')}$ is given by

$$\operatorname{cov}[Y_{j(\ell)}, Y_{j(\ell')}] = \sigma^2 v_{n_j \ell \ell'} \tag{5.8}$$

where $v_{n_j \ell \ell'}$ is the covariance of the ℓth and ℓ'th order statistics from a standard normal distribution. We assume that the samples at different conditions are independent and thus

$$\operatorname{cov}[Y_{j(\ell)}, Y_{j'(\ell')}] = 0 \quad \text{for } j \neq j' \tag{5.9}$$

We now use matrix expressions in order to derive the best linear unbiased estimators. Let

$$\underline{y} = [y_{1(1)}, \ldots, y_{1(n_{01})}, \ldots, y_{M(1)}, \ldots, y_{M(n_{0M})}]' \tag{5.10}$$

be the vector of the observed order statistics of the sample. Also denote the (k + 2) element vector of the parameters $\beta_0, \beta_1, \ldots, \beta_k$, and σ by

$$\underline{\theta} = (\underline{\beta}', \sigma)' = (\beta_0, \beta_1, \ldots, \beta_k, \sigma)' \tag{5.11}$$

The design matrix corresponding to the observed order statistics is

$$\underline{X} = \begin{bmatrix} 1 & x_{11} x_{12} & \cdots & x_{1k} z_{1,n_1} \\ & \cdots & & \cdots \\ 1 & x_{11} x_{12} & \cdots & x_{1k} z_{n_{01}, n_1} \\ & & \vdots & \\ 1 & x_{M1} x_{M2} & \cdots & x_{Mk} z_{1,n_k} \\ & \cdots & & \cdots \\ 1 & x_{M1} x_{M2} & \cdots & x_{Mk} z_{n_{0M}, n_k} \end{bmatrix} \tag{5.12}$$

Hence the regression model for the observed order statistics can be written as

$$\underline{y} = \underline{X\theta} + \epsilon \tag{5.13}$$

where $\underline{\epsilon} = \underline{y} - \underline{X\theta} = (\epsilon_{11} \cdots \epsilon_{1n_{01}} \cdots \epsilon_{M1} \cdots \epsilon_{Mn_{01}})$ denotes the error vector corresponding to y. The expectation of the error vector is the zero vector and the covariance matrix is

$$\operatorname{cov}(\epsilon, \epsilon') = \sigma^2 \underline{V} \tag{5.14}$$

where V has block diagonal form

$$\underline{V} = \operatorname{diag}(V_1, V_2, \ldots, V_M) \tag{5.15}$$

and the matrix V_j is the n_{0j} by n_{0j} matrix of the standardized order statistics of the sample at condition j, i.e., the elements are $v_{n_j \ell \ell'}$.

The regression model for censored data is thus transformed in a general linear model with correlated observations and the theory for the general linear model (see Kendall and Stuart, 1979, Vol. 2, p. 96, et seq.) can be applied to obtain estimates and variances. The best linear unbiased estimator θ_{BLU} is given by the Gauss-Markov theorem

$$\underline{\theta}_{BLU} = (\underline{X}'\underline{V}^{-1}\underline{X})^{-1}\underline{X}'\underline{V}^{-1}\underline{y} \tag{5.16}$$

provided the model is not overparameterized and the required inverses exist. The covariance matrix for θ_{BLU} is then

$$\operatorname{cov}(\underline{\theta}_{BLU}) = \sigma^2(\underline{X}'\underline{V}^{-1}\underline{X})^{-1} \tag{5.17}$$

Let ω_{ij} be the elements of this matrix; then the standard deviation of the estimators can be estimated by the unbiased estimators

$$\operatorname{STD}(\beta_{BLU,i}) = \sigma_{BLU}\sqrt{\omega_{i+1,i+1}} \qquad i = 0, \ldots, k \tag{5.18}$$

$$\operatorname{STD}(\sigma_{BLU}) = \sigma_{BLU}\sqrt{\omega_{k+2,k+2}} \tag{5.19}$$

where σ_{BLU} is the last component of the vector $\underline{\theta}_{BLU}$.

An alternative estimator for the scale parameter σ obtained from the residual sum of squares

$$SSE = (\underline{y} - \underline{X}\theta_{BLU})(y - \underline{X}\theta_{BLU})' \tag{5.20}$$

was considered by Nelson and Hahn (1973). Notice that the number of parameters to be estimated is k + 2 including σ. The sum of squares SSE has therefore $\Sigma_{j=1}^{M} n_j - (k + 2)$ degrees of freedom associated with it. The expectation of SSE is

$$E[SSE] = \left[\sum_{j=1}^{M} n_j - (k + 2)\right]\sigma^2 \tag{5.21}$$

Thus an unbiased estimator of σ^2 is

$$s^2 = SSE/\left[\sum_{j=1}^{M} n_j - (k + 2)\right] \tag{5.22}$$

This result permits alternative unbiased estimators of the dispersion matrix by substituting s^2 for σ^2 in (5.18).

Nelson and Hahn (1973) also consider the case that an estimator for N linear combinations

$$\begin{aligned} c_1 &= c_{10}\beta_0 + c_{11}\beta_1 + \cdots + c_{1k}\beta_k + c_{1k+1}\sigma \\ &\vdots \\ c_N &= c_{N0}\beta_0 + c_{N1}\beta_1 + \cdots + c_{Nk}\beta_k + c_{Mk+1}\sigma \end{aligned} \tag{5.23}$$

is required, where the coefficients c_{ij} are given. The best linear unbiased estimators for such linear combinations are obtained by substituting the best linear unbiased estimators for the model parameters into the linear combinations. The covariance matrix of these estimators $\underline{\hat{c}} = (\hat{c}_1, \ldots, \hat{c}_M)$ is

$$\operatorname{cov}(\underline{\hat{c}}) = \sigma^2 \underline{C}(\underline{X}'\underline{V}^{-1}\underline{X})^{-1}\underline{C}'$$

where $\underline{C}$ is the matrix with coefficients c_{ij} (Nelson and Hahn, 1973).

5.2 SIMPLIFIED LINEAR UNBIASED ESTIMATORS

Gupta's Linear Unbiased Estimator

The preceding results for the best linear unbiased estimators for the parameters of a linear model require that extensive tables for the elements of the matrix V be available. But these values are only tabulated for sample sizes up to n = 20. Since the exact variances and covariances of the order statistics for larger samples are unavailable, Gupta's alternative linear estimates can be used for the linear model to obtain simplified linear estimates. We substitute the covariance matrix V by the identity matrix I. Thus only the expectations $z_{i_1 n_j}$ have to be tabulated. These simplified estimators have high efficiency compared with the best linear unbiased estimators. Gupta's linear unbiased (LU) estimator is given by

$$\underline{\theta}_{LU} = (\underline{X}'\underline{X})^{-1}\underline{X}'\underline{y} \tag{5.24}$$

which is an unbiased estimator of $\underline{\theta} = (\underline{\beta}', \sigma)'$. The covariance matrix is

$$\operatorname{cov}(\underline{\theta}_{LU}) = \sigma^2(\underline{X}'\underline{X})^{-1}\underline{X}'\underline{V}\underline{X}(\underline{X}'\underline{X})^{-1} \tag{5.25}$$

We notice that the last column of the matrix $\underline{X}$ (5.12) was defined by the expected order statistics of a standard normal sample and therefore only these values must be known to calculate $\underline{\theta}_{LU}$.

These expected order statistics are tabulated by Harter (1969) for sample sizes up to 400. In order to obtain the covariance matrix of the estimator $\hat{\underline{\beta}}$ for n larger than 20, approximations suggested by Prescott (1970) may be used. It was also shown by Ali and Chan (1964) that Gupta's estimators are asymptotically efficient and normally distributed.

Nelson and Hahn's Linear Unbiased Estimator

Nelson and Hahn (1973) suggested a regression based on the estimates for the means at the M test conditions rather than on the original data. For estimating the means at the M test conditions they suggest using the BLU estimator described in Section 4.3.

The estimators $\mu_{BLU,j}$ (j = 1, 2, ..., M) are independent, since the samples at each condition were assumed to be independent. Hence the covariance matrix of the vector $\underline{\mu}_{BLU} = (\mu_{BLU,1}, \ldots, \mu_{BLU,M})$ is diagonal

$$\underset{\sim}{\underline{\tilde{V}}} = \sigma^2 \begin{bmatrix} \mathrm{var}(\mu_{\mathrm{BLU},1}) & 0 \\ 0 & \mathrm{var}(\mu_{\mathrm{BLU},M}) \end{bmatrix} \tag{5.26}$$

The design matrix is obtained from the matrix X by deleting the last column. Thus let

$$\underline{\tilde{X}} = \begin{bmatrix} 1 & x_{11} x_{12} & \cdots & x_{1k} \\ 1 & x_{21} x_{22} & \cdots & x_{2k} \\ 1 & x_{M2} x_{k2} & \cdots & x_{MK} \end{bmatrix} \tag{5.27}$$

be the new design matrix and define $\underline{\tilde{\theta}} = (\beta_0, \beta_1, \ldots, \beta_k)'$. Note that the previous defined parameter vector was $\underline{\theta} = (\underline{\beta}', \sigma)'$. Instead of the original observation y, the means at each condition are used to perform a linear regression. The model for μ_{BLU} is therefore given by

$$\underline{\mu}_{\mathrm{BLU}} = \underline{\tilde{X}}\underline{\beta} + \epsilon \tag{5.28}$$

The vector $\underline{\epsilon} = \underline{\mu}_{\mathrm{BLU}} - \underline{\tilde{X}}\underline{\beta}$ has zero expectation and variance $V_{\mu_{\mathrm{BLU}}}$. The best linear unbiased estimator is (Nelson and Hahn, 1973)

$$\underline{\tilde{\beta}}_{\mathrm{SLU}} = (\underline{\tilde{X}}'\underline{\tilde{V}}^{-1}\underline{\tilde{X}})^{-1}\underline{\tilde{X}}'\underline{\tilde{V}}^{-1}\underline{\mu}_{\mathrm{BLU}} \tag{5.29}$$

It is assumed again that the model is not overparameterized and that the required matrix inverses exist. The covariance matrix of $\tilde{\beta}_{\mathrm{BLU}}$ is

$$\mathrm{cov}(\underline{\tilde{\beta}}_{\mathrm{SLU}}) = \sigma^2(\underline{\tilde{X}}'\underline{\tilde{V}}^{-1}\underline{\tilde{X}})^{-1} \tag{5.30}$$

Note that the estimator $\underline{\tilde{\beta}}_{\mathrm{SLU}}$ is unbiased but it is less efficient than the unbiased estimator based on the original observations. The best linear unbiased estimator of σ based on $\tilde{\sigma}_1, \ldots, \tilde{\sigma}_M$, the unbiased estimators of σ at each of the test conditions, are given by the weighted average

$$\tilde{\sigma} = \mathrm{var}(\tilde{\sigma}) \sum_{j=1}^{M} [\tilde{\sigma}_j/\mathrm{var}(\tilde{\sigma}_j)] \tag{5.31}$$

where

$$\mathrm{var}(\tilde{\sigma}) = \frac{1}{1/[\mathrm{var}(\tilde{\sigma}_1)] + \cdots + 1/[\mathrm{var}(\tilde{\sigma}_M)]} \tag{5.32}$$

is the variance of $\tilde{\sigma}$ and var $(\tilde{\sigma}j)$ is the variance of $\tilde{\sigma}$ based on the replications at the jth test condition (Nelson and Hahn, 1973).

For the case where the sample size and the number of censored observations are the same at each test condition, Nelson and Hahn (1973) give a simplified version of their estimator. Consider M test conditions each having n observations, n_r, of which are censored. Then the variances of the M estimators are all identical.

Hence, the best linear unbiased estimator $\tilde{\underline{\beta}}_{SLU}$ based on the unbiased estimates $\mu_{BLU,1}, \ldots, \mu_{BLU,M}$ is derived from (5.29)

$$\tilde{\underline{\beta}}_{SLU} = (\tilde{\underline{X}}'\tilde{\underline{X}})^{-1}\tilde{\underline{X}}'\underline{\mu}_{BLU} \tag{5.33}$$

This estimator might also be used in the more general case of censoring, but with some loss of efficiency. We notice that this estimator is still unbiased even in the general case of censoring.

5.3 MAXIMUM LIKELIHOOD ESTIMATOR

Type I Censored Samples

Consider the linear model

$$Y_j = \underline{\beta}'\underline{x}_j + \epsilon \qquad j = 1, \ldots, M \tag{5.35}$$

where ϵ is a normally distributed random variable with mean zero and variance σ^2. The parameter vector $\underline{\beta}$ has dimension k + 1. Suppose there are M test conditions and the sample size at the jth condition is n_j. Define $\mu_j = \Sigma_{i=0}^{k} \beta_i x_{ij}$ with $x_{0j} \equiv 1$. The samples are assumed to be Type I censored from above (on the right), where n_{rj} observations are available at the jth condition. Let y_{rj} be the censoring points and let

$$y_{j(1)} \leqslant y_{j(2)} \leqslant \cdots \leqslant y_{j(n_{0j})}$$

be the order statistics of this sample. The likelihood function of the M samples is proportional to

$$L(\beta, \sigma) \propto \frac{1}{\sigma^{n_0}} \prod_{j=1}^{M} \left\{ \left[1 - \Phi\left(\frac{y_{rj} - \mu_j}{\sigma} \right) \right]^{n_{rj}} \prod_{\ell=1}^{n_{0j}} \phi \left[\frac{y_{j(\ell)} - \mu_j}{\sigma} \right] \right\} \tag{5.36}$$

where $n_0 = \Sigma_{j=1}^{M} n_{0j}$.

The natural logarithm of the likelihood function is thus up to a constant

$$\log L = -n_0 \log \sigma + \sum_{j=1}^{M} n_{rj} \log \left[1 - \Phi\left(\frac{y_{rj} - \mu_j}{\sigma} \right) \right] - \frac{1}{2} \sum_{j=1}^{M} \sum_{\ell=1}^{n_{0j}} \frac{(y_{j(\ell)} - \mu_j)^2}{\sigma^2} \tag{5.37}$$

The partial derivatives of the log-likelihood function are

$$\frac{\partial \log L}{\partial \beta_i} = \frac{1}{\sigma} \sum_{j=1}^{M} n_{rj} \frac{\phi(u_{rj})}{1 - \Phi(u_{rj})} x_{ij} + \frac{1}{\sigma^2} \sum_{j=1}^{M} \sum_{\ell=1}^{n_{0j}} (y_{j(\ell)} - \mu_j) x_{ij}$$

$$i = 0, 1, \ldots, k \tag{5.38}$$

$$\frac{\partial \log L}{\partial \sigma} = -\frac{n_0}{\sigma} + \frac{1}{\sigma} \sum_{j=1}^{M} n_{rj} \frac{\phi(u_{rj})}{1 - \Phi(u_{rj})} u_{rj} + \frac{1}{\sigma^3} \sum_{j=1}^{M} \sum_{\ell=1}^{n_{0j}} (y_{j(\ell)} - \mu_j)^2 \tag{5.39}$$

where $u_{rj} = (y_{rj} - \mu_j)/\sigma$.

In order to obtain the MLEs we set (5.38) and (5.39) equal to zero and solve the equations using the Newton-Raphson method. This method will have quadratic convergence given that accurate starting values are used. Good starting values can be obtained by applying one of the simplified methods described in the next section.

An alternate method of computing maximum likelihood estimates was suggested by Dempster et al. (1977). Their expectation maximization (EM) algorithm was generally derived for parametric models, but the idea was also applied to a nonparametric model by Buckley and James (1979) and Schneider and Weissfeld (1986). Aitkin (1981) showed how the EM algorithm is applied to the normal censored linear model.

The EM algorithm requires on the average more iterations to convergence than the Newton-Raphson method. But the EM algorithm is much easier to program and gives also some understanding of the estimation problem based on censored data. In what follows we shall present the EM algorithm following Aitkin (1981).

To solve the equations $\partial \log L/\partial\beta_i = 0$ and $\partial \log L/\partial\sigma = 0$ we can rewrite (5.38) and (5.39) to obtain the k + 2 equations

$$\sum_{j=1}^{M}\left[\sigma n_{rj}\tilde{W}(u_{rj})x_{ij} + \sum_{\ell=1}^{n_{0j}}(y_{j(\ell)} - \mu_j)x_{ij}\right] = 0 \qquad i = 0, 1, \ldots, k \tag{5.40}$$

and

$$\left[n_0 - \sum_{j=1}^{M} n_{rj}\tilde{W}(u_{rj})u_{rj}\right]\left[\frac{\sum_{j=1}^{M}\sum_{\ell=1}^{n_{0j}}(y_{j(\ell)} - \mu_j)^2}{n_0 - \sum_{j=1}^{M} n_{rj}\tilde{W}(u_{rj})u_{rj}} - \sigma^2\right] = 0 \tag{5.41}$$

where $\tilde{W}(u_{rj}) = \phi(u_{rj})/[1 - \Phi(u_{rj})]$.

The expected value of the censored observations at the jth test condition is given by

$$E[Y_j | Y_j > y_{rj}] = \mu_j + \sigma\tilde{W}(u_{rj}) \tag{5.42}$$

To show this we note that the conditional distribution of y is

$$F(y \mid y > y_r) = \int_{y_r}^{y} \frac{\phi[(x - \mu)/\sigma]/\sigma}{1 - \Phi(u_r)} \, dx$$

and

$$E[y \mid y > y_r] = \frac{1}{1 - \Phi(u_r)} \int_{u_r}^{\infty} (\sigma x + \mu)\phi(x) \, dx$$

Straightforward integration then yields the result given above. We may therefore rewrite the equations (5.40)

$$\sum_{j=1}^{M} \sum_{\ell=1}^{n_j} (\tilde{y}_{j(\ell)} - \mu_j)x_{ij} = 0 \qquad i = 0, 1, \ldots, k \tag{5.43}$$

where

$$\tilde{y}_j = \delta y_{j(\ell)} + (1 - \delta)[\mu_j + \sigma\tilde{W}(u_{rj})] \tag{5.44}$$

and

$$\delta = \begin{cases} 1 & \text{if } \ell \leq n_{0j} \\ 0 & \text{if } n_{0j} < \ell \leq n_j \end{cases} \tag{5.45}$$

We also note that

$$E[y_j^2 \mid y_j > y_{rj}] = \mu_j^2 + \sigma^2 + \sigma(y_{rj} + \mu_j)\tilde{W}(u_{rj}) \tag{5.46}$$

which is given by straightforward integration of

$$E[y_j^2 \mid y_j > y_{rj}] = \frac{1}{1 - \Phi(u_{rj})} \int_{u_{rj}}^{\infty} (\sigma x + \mu_j)^2 \phi(x) \, dx$$

Let

$$\epsilon^2_{j(\ell)} = y^2_{j(\ell)} - 2y_{j(\ell)}\mu_j + \mu^2_j = (y_{j(\ell)} - \mu_j)^2 \tag{5.47}$$

then the equation (5.41) can be written as

$$n\sigma^2 - \sum_{j=1}^{M} \sum_{\ell=1}^{n_j} \tilde{\epsilon}^2_{j(\ell)} = 0 \tag{5.48}$$

where $n = \Sigma_{j=1}^{M} n_j$, and with (5.45) we obtain

$$\tilde{\epsilon}^2_{j(\ell)} = \delta(y_{j(\ell)} - \mu_j)^2 + (1 - \delta)\Big\{E[y^2_j \mid y_j > y_{rj}]$$

$$- 2\mu_j E[y_j \mid y_j > y_r] + \mu^2_j\Big\} \tag{5.49}$$

The equations (5.43) and (5.48) have the same form as the maximum likelihood equations for uncensored data, where the censored observations are replaced by their expected values. But notice that in the formula for estimation of the variance σ^2, y^2_j has to be substituted by $E[y^2_j \mid y_j \geqslant y_r]$ rather than by $E[y_j \mid y_j > y_{rj}]^2$. We write the latter result in matrix notation. Let

$$\underline{B} = \begin{bmatrix} 1 & x_{11} x_{12} \cdots x_{1k} \\ 1 & \\ 1 & x_{1k} \\ \vdots & \\ 1 & x_{Mk} \\ 1 & \\ 1 & x_{Mk} \end{bmatrix} \tag{5.50}$$

be the design matrix with $k + 1$ columns and $n = \Sigma_{j=1}^{M} n_j$ rows. Then the maximum likelihood estimator of β is identical with the ordinary least square estimator

$$\underline{\beta}_{ML} = (\underline{B}'\underline{B})^{-1}\underline{B}'\underline{\tilde{y}} \tag{5.51}$$

where the censored observations are replaced by their expected values. Furthermore, the maximum likelihood estimator for σ^2 is given by the sum of squared residuals

$$\sigma^2_{ML} = \frac{1}{n}(\underline{\tilde{y}} - \underline{B}\underline{\beta}_{ML})'(\underline{\tilde{y}} - \underline{B}\underline{\beta}_{ML}) \tag{5.52}$$

where the censored observations are replaced by their expectations and squares of censored observations are replaced by their second moments.

This is, of course, not an explicit result. The estimation of $\underline{\beta}_{ML}$ and σ^2_{ML} has to be done simultaneously since the expected values of censored observations depend on $\underline{\beta}$ and on σ^2.

The maximum likelihood estimators $\underline{\beta}_{ML}$ and σ^2_{ML} can be found by iteration, where in every step new estimates of the expectations $E[y \mid y > y_r]$ and $E[y^2 \mid y > y_r]$ will be used to calculate a new estimate $\underline{\beta}_{ML}$ and σ^2_{ML}. The algorithm known as the EM algorithm (Dempster, 1977) and can be described as follows (see also Aitkin, 1981):

Step 1: Find initial estimates of the expected value of the censored observations and prepare the vector $y^{(0)}$.

Step 2: Calculate for i = 1, 2, ...

$$\underline{\beta}^{(i)}_{ML} = (\underline{B}'\underline{B})^{-1}\underline{B}'\underline{\tilde{y}}^{(i-1)}$$

and

$$\sigma^{2(i)}_{ML} = \frac{1}{n}[\underline{y}^{(i-1)} - \underline{B}\underline{\beta}^{(i)}_{ML}][\underline{y}^{(i-1)} - \underline{B}\underline{\beta}^{(i)}_{ML}]'$$

until the estimates are accurate enough for practical purposes. Since the bias of the ML estimator of σ in the ordinary regression case can be removed by a degree of freedom adjustment Aitkin (1981) suggested using the bias corrected estimate $\hat{\sigma}^2 = \sigma^2_{ML}\, n_0/(n_0 - k - 1)$. This procedure indeed reduces the bias in σ significantly, as will be shown in Section 5.5.

A computer program for calculating the maximum likelihood estimates via the EM algorithm is given by Wolynetz (1979).

Asymptotic variance. Amemiya (1973) showed that under weak assumptions, which are usually fulfilled in practical situations, the likelihood equations have a strongly consistent root and $\sqrt{n_0}\,(\theta_{ML} - \theta_0)$ is asymptotically normally distributed with mean zero and covariance given by the inverse of the Fisher information matrix. The proof is given by Amemiya (1973).

Type I censoring. Notice that the second derivatives of the log-likelihood function for Type I censored samples are

$$\frac{\partial^2 \log L}{\partial\beta_i\,\partial\beta_{i'}} = \frac{1}{\sigma^2}\sum_{j=1}^{M}[n_{rj}u_{rj}\widetilde{W}(u_{rj}) - n_{rj}\widetilde{W}^2(u_{rj}) - n_{0j}]x_{ij}x_{i'j} \tag{5.53}$$

$$\frac{\partial^2 \log L}{\partial\sigma\partial\beta_i} = \frac{1}{\sigma^2}\sum_{j=1}^{M}\Bigg\{n_{rj}[u_{rj}\widetilde{W}(u_{rj})(u_{rj} - \widetilde{W}(u_{rj})) - \widetilde{W}(u_{rj})] - 2\sum_{\ell=1}^{n_{0j}}(Y_{j(\ell)} - \mu_j)/\sigma\Bigg\}x_{ij} \tag{5.54}$$

$$\frac{\partial^2 \log L}{\partial\sigma\partial\sigma} = \frac{n_0}{\sigma^2} - \frac{1}{\sigma^2}\sum_{j=1}^{M}n_{rj}u_{rj}\widetilde{W}(u_{rj})[2 + u_{rj}(\widetilde{W}(u_{rj}) - u_{rj})] - \frac{3}{\sigma^4}\sum_{j=1}^{M}\sum_{\ell=1}^{n_{0j}}(Y_{j(\ell)} - \mu_j)^2 \tag{5.55}$$

Taking expectations we obtain the elements of the information matrix which are given as

$$J = \begin{bmatrix} \sum_{j=1}^{M} n_j a_j x_{0j} x_{0j} & \cdots & \sum_{j=1}^{M} n_j a_j x_{0j} x_{kj}, & \sum_{j=1}^{M} n_j b_j x_{0j} \\ \vdots & & \vdots & \\ \sum_{j=1}^{M} n_j a_j x_{kj} x_{0j} & \cdots & \sum_{j=1}^{M} n_j x_{kj} x_{kj}, & \sum_{j=1}^{M} n_j b_j x_{kj} \\ \sum_{j=1}^{M} n_j b_j x_{0j} & \cdots & \sum_{j=1}^{M} n_j b_j x_{kj}, & \sum_{j=1}^{M} n_j c_j \end{bmatrix} \tag{5.56}$$

where

$$a_j = \frac{1}{\sigma^2} \Phi(u_{rj}) \{1 + \widetilde{W}(-u_{rj})[\widetilde{W}(u_{rj}) - u_{rj}]\} \tag{5.57}$$

$$b_j = -\frac{1}{\sigma^2} \Phi(\Phi_{rj}) \widetilde{W}(-u_{rj}) \{1 - u_{rj}[\widetilde{W}(u_{rj}) - u_{rj}]\} \tag{5.58}$$

$$c_j = \frac{1}{\sigma^2} [2\Phi(u_{rj}) + u_{rj} b_j] \tag{5.59}$$

Notice that σ^2 and $\underline{\beta}$ are the true parameters. We may estimate $\underline{J}$ simply by replacing the true values with estimates.

The information matrix is a straightforward generalization of the information matrix derived for the univariate model.

Type II censored samples. We notice that the maximum likelihood estimation procedure is the same for Type I and Type II censored samples. We simply substitute the largest order statistics $y_{j(n_{0j})}$ for the truncation points y_{rj}. We notice, however, that the variance of the estimators for Type I and Type II censoring is in principle different. But, as mentioned in Chapter 4, this difference diminishes asymptotically.

Asymptotic variance. The elements of the information matrix alter according to (4.31)-(4.33). For Type II censored samples we thus obtain

$$a_j = \frac{1}{\sigma^2}\left\{1 + \frac{n_{rj}}{n_{0j}} \tilde{W}(u_{rj})[\tilde{W}(u_{rj}) - u_{rj}]\right\} n_{0j}/n_j \tag{5.60}$$

$$b_j = -\frac{1}{\sigma^2}\frac{n_{rj}}{n_{0j}} \tilde{W}(u_{rj})\left\{1 - u_{rj}[\tilde{W}(u_{rj}) - u_{rj}]\right\} n_{0j}/n_j \tag{5.61}$$

$$c_j = \frac{1}{\sigma^2}[2 + u_{rj}b_j]n_{0j}/n_j \tag{5.62}$$

where $u_{rj} = \Phi^{-1}(n_{0j}/n_j)$. We notice that for Type II censoring the covariance matrix is known, while for Type II censoring the covariance matrix has to be estimated. Type II censoring therefore offers some advantage over Type I censoring in that the confidence intervals are generally more reliable. But Type I censoring seems to be preferred in praxis due to its practical advantage of being able to plan the length of an experiment in advance.

5.4 SIMPLIFIED ESTIMATORS

In this section we describe simplified techniques to estimate $\underline{\beta}$ and σ in the censored linear model. The iterative least square method proposed by Schmee and Hahn (1979) is an iterative procedure like the maximum likelihood method, but it uses a more convenient estimator for σ^2. The modified maximum likelihood (MML) estimator (Tiku, 1978) and the restricted maximum likelihood (RML) estimator described in Section 4.4 can be applied only in a model with replicates at each test condition, i.e., $n_j \geq 2$, $j = 1, \ldots, M$. Furthermore, at least one uncensored observation has to be available at each test condition, i.e., $n_{0j} \geq 1$, $j = 1, \ldots, M$. If the latter assumptions are fulfilled both estimators MML and RML have high efficiency. The iterative least square estimator β_{ILS} of Schmee and Hahn has high efficiency for light to moderate censoring despite the fact that σ_{ILS} is heavily biased and has low efficiency for severe censoring.

Iterative Least Squares

The iterative least square method proposed by Schmee and Hahn (1979) follows the same procedure to find an estimator of β as the maximum likelihood method. However, the estimator of σ^2 suggested by Schmee and Hahn differs from the maximum likelihood estimator. If we substitute $E[y_j | y_j > y_{rj}]^2$ for $E[y_j^2 | y_j > y_{rj}]$ we obtain

$$n\sigma^2 = -\sum_{j=1}^{M} \sum_{\ell=1}^{n_{0j}} (y_{i(\ell)} - \mu_i)^2 - \sigma^2 \sum_{j=1}^{M} n_{rj} \tilde{W}(u_{rj}) \qquad (5.63)$$

Because the bias of the ML-estimator of σ^2 in the ordinary regression case can be removed by a degree of freedom adjustment Schmee and Hahn suggested the iterative estimator

$$(n - k - 1)\sigma^{2(i)} = \sum_{j=1}^{M} \sum_{\ell=1}^{n_{0j}} (y_{j(\ell)} - \mu_j)^2 + \sigma^{2(i-1)} \sum_{j=1}^{M} n_{rj} \tilde{W}(u_{rj}) \qquad (5.64)$$

But we notice that despite the degree of freedom adjustment this estimator is heavily biased while the adjusted ML-estimator is nearly unbiased.

A computer program for calculating the iterative least squares estimates $\underline{\beta}_{ILS}$ and σ_{ILS} is given by Schmee and Hahn (1981).

Tiku's Simplified Estimator

Simplified estimators for β and σ were suggested by Tiku (1978). His estimators are based on the same principle applied to the univariate model in Chapter 4. The hazard rate $\tilde{W}(u)$ is approximated by a linear function which results in explicit estimates $\hat{\underline{\beta}}$ and $\hat{\sigma}^2$.

The asymptotic properties of these estimators are the same as those of the maximum likelihood estimators since both are asymptotically identical. Tiku (1978) investigates the special case of symmetrical Type II censored samples. But this procedure can easily be generalized.

The formal steps to obtain Tiku's modified maximum likelihood estimation are the same as described in Section 4.4. For a detailed description of the MML estimator we refer to Tiku (1978). We notice that at least one observation has to be available at each design point x_j in order to apply Tiku's estimator.

Restricted Maximum Likelihood Estimator

We shall derive a restricted maximum likelihood estimator which is an extension of the estimator, described in Section 4.4, estimating the parameters of a linear model. We restrict the discussion to the case of singly Type I right censored samples but mention that the extension to doubly censored samples or Type II censored samples is straightforward.

Consider the linear model

$$y_j = \underline{\beta}'\underline{x}_j + \epsilon_j \qquad j = 1, \ldots, M$$

where ϵ_j is a normally distributed random variable with mean zero and variance σ^2. The parameter vector β has dimension $k + 1$. Suppose there are M test conditions and the sample size at the jth condition is $n_j > 1$. Define $\mu_j = \Sigma_{i=0}^{k} \beta_i x_{ij}$ with $x_{0j} \equiv 1$. The samples are assumed to be censored from above (on the right), where n_{rj} observations of the sample at the jth condition are censored, i.e., lie above a preassigned value y_{rj}. Thus $n_{0j} = n_j - n_{rj} > 0$ of the n_j observations are assumed to be available at the jth condition. Let

$$y_{j(1)} \leqslant y_{j(2)} \leqslant \cdots \leqslant y_{j(n_{0j})}$$

be the order statistics of this sample. The likelihood function of the samples is proportional to

$$L(\beta, \sigma) \propto \frac{1}{\sigma^{n_0}} \prod_{j=1}^{M} \left\{ \left[1 - \Phi\left(\frac{y_{rj} - \mu_j}{\sigma}\right)\right]^{n_{rj}} \sum_{\ell=1}^{n_{0j}} \phi\left(\frac{y_{j(\ell)} - \mu_j}{\sigma}\right) \right\}$$

where $n_0 = \Sigma_{j=1}^{M} n_{0j}$.

The natural logarithm of the likelihood function is thus up to a constant

$$\log L = -n_0 \log \sigma + \sum_{j=1}^{M} n_{rj} \log \left[1 - \Phi\left(\frac{y_{rj} - \mu_j}{\sigma}\right)\right] - \frac{1}{2} \sum_{j=1}^{M} \sum_{\ell=1}^{n_{0j}} \frac{(y_{j(\ell)} - \mu_j)^2}{\sigma^2} \qquad (5.65)$$

We introduce the restriction $\Phi(u_{rj}) = 1 - n_{rj}/n_j$ where $u_{rj} = (y_{rj} - \mu_j)/\sigma$; thus the log-likelihood function (5.65) can be written (up to a constant term) as

$$\log L = -n_0 \log \sigma - \frac{1}{2} \sum_{j=1}^{M} \sum_{=1}^{n_{0j}} (-y^*_{j(\ell)}/\sigma + u_{rj})^2 \tag{5.66}$$

where $y^*_{j(\ell)} = y_{jr} - y_{j(\ell)}$. Since u_{rj} is fixed we obtain an estimator of σ from $\partial \log L/\partial \sigma = 0$. Thus

$$\frac{\partial \log L}{\partial \sigma} = -\frac{n_0}{\sigma} - \frac{1}{\sigma^2} \sum_{j=1}^{M} \sum_{\ell=1}^{n_{0j}} (-y^*_{j(\ell)}/\sigma + u_{rj})y^*_{j(\ell)} = 0 \tag{5.67}$$

and the restricted maximum likelihood estimator can be written as

$$\sigma_{RML} = \frac{1}{2} [-\theta + (\theta^2 + 4v)^{1/2}] \tag{5.68}$$

where

$$\theta = \frac{1}{n_0} \sum_{j=1}^{M} \Phi^{-1}(1 - n_{rj}/n_j) \sum_{\ell=1}^{n_{0j}} (y_{rj} - y_{j(\ell)}] \tag{5.69}$$

and

$$v = \sum_{j=1}^{M} \sum_{\ell=1}^{n_{0j}} [y_{rj} - y_{j(\ell)}]^2/n_0 \tag{5.70}$$

This estimator of σ is consistent and asymptotically normally distributed as shown by Persson and Rootzen (1977). Since this estimator is biased for small sample sizes we propose here the bias correction also suggested by Tiku (1978) and Aitkin (1981):

$$\sigma'_{RML} = \sigma_{RML}\sqrt{n_0/(n_0 - k - 1)} \tag{5.71}$$

Our aim is now to estimate the expected values of the censored observations. Having these estimates $\hat{y}_j$, we define

$$\tilde{y}_{j,\ell} = \delta y_{j,\ell} + (1 - \delta)\hat{y}_{j,\ell} \tag{5.72}$$

where

$$\delta = \begin{cases} 1 & \text{if } y_{j,} \text{ is not censored} \\ 0 & \text{if } y_{j,} \text{ is censored} \end{cases}$$

Let

$$\overline{y}_j = \sum_{\ell=1}^{n_{0j}} y_{j\ell}/n_{0j}$$

then applying (2.23) and (5.42) we obtain an estimator of $E[y_j | y_j > y_{rj}]$

$$\hat{y}_j = \overline{y}_j + \sigma'_{RML}\phi[\Phi^{-1}(1 - n_{rj}/n_j)] \frac{n_j^2}{n_{0j}n_{rj}} \tag{5.73}$$

The ordinary least squares estimate

$$\beta_{RML} = (\underline{B}'\underline{B})^{-1}\underline{B}'\tilde{y} \tag{5.74}$$

where y is defined by (5.72) and $\underline{B}$ by (5.50), is denoted as restricted maximum likelihood estimator. The estimator $\underline{\beta}_{RML}$ can now be used to obtain an alternative estimate $\hat{\sigma} = \sqrt{SSE/(n - k - 1)}$. But the latter estimator of σ is not necessarily more efficient than σ'_{RML} as simulation results show.

Simplified Restricted Maximum Likelihood Estimator

We also consider a simplification of the RML-estimator β_{RML}. The procedure described below is similar to the Nelson Hahn estimator described in Section 5.2. Instead of using the original data $y_{j\ell}$ in (5.74) we replace the n_{0j} equations at each test condition by one equation, namely $\hat{y}_j = \hat{\beta}'x_j$, where $\hat{y}_j$ is the estimated mean at the jth condition, i.e., $\hat{y}_j = \bar{y}_j + \sigma'_{RML}W(u_{rj})$. As a result of this procedure we obtain the M by (k + 1) design matrix

$$\underline{c} = \begin{bmatrix} 1 & x_{11}x_{12} & \cdots & x_{1k} \\ \vdots & & & \\ 1 & x_{1M}x_{2M} & \cdots & x_{Mk} \end{bmatrix} \tag{5.75}$$

Provided this model is not overparameterized we obtain the estimator

$$\beta^r_{RML} = (\underline{c}'\underline{c})^{-1}\underline{c}'\hat{y} \tag{5.76}$$

We notice that there is some loss of efficiency when the expected values instead of the original data are used.

Total Censoring at One Condition

The above described procedure is only applicable when at every test condition there is at least one uncensored observation left. If at some test conditions all observations are censored we suggest a two-step algorithm.

Step 1. Calculate the RML-estimates σ_{RML} and $\underline{\beta}^{(1)}_{RML}$ using only such test conditions where at least one observation is available. Evaluate $\hat{y}_{j'} = \underline{\beta}^{(1)}_{RML}\underline{x}_{j'}$ for each condition j' with no observations left. The estimate $\tilde{y}_{j'} = \hat{y}_{j'} + \sigma_{RML}\tilde{W}(u_{rj})$ is now used to represent the censored observations at this condition where $u_{rj'}$ is now given by $u_{rj} = (y_{rj} - \hat{y}_j)/y_{j'}\sigma_{RML}$.

Step 2. Calculate the RML-estimate $\underline{\beta}^{(2)}_{RML}$ using $\tilde{y}_{j'}$ as observations at the j'th condition.

There are several computer programs for computing estimates in the censored linear model available. They are discussed in Appendix B.

5.5 PROPERTIES OF THE ESTIMATORS

In this section we shall compare three estimators presented in Chapter 5 by means of a Monte Carlo simulation study. Furthermore we shall discuss the robustness of the estimators in the case of heteroscedasticity. An example presented by Crawford (1970) is reconsidered to demonstrate some important problems in application. These problems concern outliers and total censoring.

We did not include Tiku's MML estimator in the following study; we only mention that this estimator gives estimates which are close to likelihood estimates (see Tiku, 1978), provided there is at least one uncensored observation for each test condition.

Monte Carlo Simulations

We consider the linear model

$$y_{j,i} = \beta_0 + \beta_1 x_j + \epsilon_{j,i} \qquad j = 1, 2, 3, 4, 5 \qquad i = 1, \ldots, n_j$$

where without loss of generality $\epsilon_j \sim N(0, 1)$ and $x_j = j$, $\beta_0 = 1$, $\beta_1 = 1$. The number of replications n_j at each condition was chosen to be equal, i.e., $n_1 = n_2 = n_3 = n_4 = n_5$. The censoring scheme was Type I censoring with 1) equal censoring and 2) increasing censoring with j. The degree of censoring is given as the average percentage of observations censored at each test condition. The bias in percent of the standard deviation and the mean square error are used as measures of performance. The simulation results are based on 10,000 samples, where the IMSL subroutine GGNML was used to generate normal variates. Table 5.1 shows the results for equal censoring. As mentioned in Section 5.3, the bias of the MLE $\hat{\sigma}$ can be reduced by a degree of freedom adjustment. But this can be done in two different ways, which results in different estimates of $\underline{\beta}$. First, we can use the adjusted estimator of the standard deviation $\hat{\sigma}$ during the iteration for computing $\hat{\underline{\beta}}$. Second, we simply define $\hat{\sigma}^2_{MLa} = \hat{\sigma}^2_{ML} n_0/(n_0 - k - 1)$, where σ_{ML} is the final MLE. The first method yields an estimator of σ which is slightly less biased than the one obtained by the second method but results in higher variance of $\hat{\underline{\beta}}$. We have used the second method of adjusting $\hat{\sigma}_{ML}$.

For light censoring all three estimators (IML, ILS, and RML) perform equally well. For severe censoring the RML-estimators behave like the IML-estimators. The ILS estimator of $\underline{\beta}_0$ is heavily

TABLE 5.1 Simulation Results for Equal Censoring

n_j	10						20						
	$\hat{\beta}_0$		$\hat{\beta}_1$		$\hat{\sigma}$		$\hat{\beta}_0$		$\hat{\beta}_1$		$\hat{\sigma}$		Censoring
	Bias	MSE	Bias	MSE	Bias	MSE	Bias	MSE	Bias	MSE	Bias	MSE	(%)
ILS	0	117	0	11	-3	13	0	55	0	5	-3	7	
IML	0	118	0	11	1	15	0	56	0	5	0	7	20
RML	-2	123	0	11	-1	14	-1	58	0	5	-2	8	
ILS	-3	118	0	11	-10	24	-3	56	0	5	-10	17	
IML	1	132	0	12	1	22	1	63	0	6	0	10	40
RML	-2	146	0	13	-1	21	-1	67	0	6	-1	10	
ILS	-16	130	0	11	-26	83	-15	73	0	5	-25	72	
IML	1	165	0	15	3	37	1	83	0	7	0	17	60
RML	-3	199	0	17	0	35	0	93	0	8	0	17	

Note: MSE has to be multiplied by 0.001. Bias in percent of σ, 0 stands for less than 1%.

biased when censoring becomes severe at each test condition. The RML- and the IML-estimators of β_0 and β_1 are nearly unbiased for all degrees of censoring. But we notice that the MSE of the ILS-estimator is usually smaller than the MSE of the IML- and RML-estimator. The ILS-estimator of σ becomes unreliable for severe censoring; it is heavily biased and has an extremely large MSE.

For unequal censoring (see Table 5.2) the ILS estimators β_{ILS} and σ_{ILS} become significantly biased, whereas the IML- and RML-estimators are almost unbiased but $\underline{\beta}_{IML}$ and $\underline{\beta}_{RML}$ usually have higher MSE than $\underline{\beta}_{ILS}$. This result, of course, only holds for small sample sizes, since the IML-estimators are asymptotically efficient. But even for sample sizes $n_j = 20$ the ILS-estimator $\underline{\beta}_{ILS}$ has surprisingly small MSE in comparison to the IML.

Schmee and Hann (1979) presented a simulation study for a design without replications. Their results are basically the same, though they do not reveal the poor quality of the σ_{ILS} estimator.

Effect of Heteroscedasticity

Frequently the assumption of homoscedasticity does not hold. The linear model can then be written as

$$y_{ij} = \beta' x_j + \sigma_j \epsilon_{ij} \qquad i = 1, \ldots, n_j \qquad j = 1, \ldots, M \tag{5.77}$$

where y_{ij} is the response of the ith replicate at the design point x_j. The standard deviations σ_j express the heteroscedasticity in the model and the ϵ_{ij} are identically distributed normal random variables with mean zero and variance 1.

The uncensored heteroscedastic linear model is the focus of some attention in literature; see Carroll and Ruppert (1982). Weighted least squares as considered by Bement and Williams (1969), and Fuller and Rao (1978) is a straightforward simple technique to obtain an estimator of $\underline{\beta}$. The weighted least square estimator $\underline{\beta}_{WLS}$ is obtained in two steps. The first step is to obtain σ_j, $j = 1, \ldots, M$. The second step is a weighted least squares estimation. Unfortunately, in practical cases where n_j is small, this estimator usually has higher variance than the ordinary least squares estimator β_{LS} (see Fuller and Rao, 1978, and Jacquez et al., 1968). Some authors (see Carroll and Ruppert, 1982) therefore consider a model where the variances σ_j can be modeled as

$$\sigma_j = H(x_j, \theta) \tag{5.78}$$

TABLE 5.2 Simulation Results for Increasing Censoring

n_j	5				10				20			
	$\hat{\beta}_0$		$\hat{\beta}_1$		$\hat{\beta}_0$		$\hat{\beta}_1$		$\hat{\beta}_0$		$\hat{\beta}_1$	
	Bias	MSE	Bias	MSE	Bias	MSE	Bias	MSE	Bias	MSE	Bias	MSE
ILS	4	130	5	18	21	57	5	63	-6	10	21	49
IML	-2	143	1	20	0	25	-1	67	0	9	0	12
RML	1	153	-1	22	2	26	0	79	0	12	0	13

<u>Note</u>: MSE has to be multiplied by 0.001. Bias in percent of σ, 0 stands for less than 1%.

TABLE 5.3 Simulation Results for Heteroscedastic Model

n_j	5				10				20			
	$\hat{\beta}_0$		$\hat{\beta}_1$		$\hat{\beta}_0$		$\hat{\beta}_1$		$\hat{\beta}_0$		$\hat{\beta}_1$	
	Bias	MSE	Bias	MSE	Bias	MSE	Bias	MSE	Bias	MSE	Bias	MSE
ILS	14	426	-5	49	15	216	-6	25	15	115	-6	14
IML	17	481	-5	52	18	245	-6	26	18	133	-6	15
RML	1	493	1	68	0	225	1	31	1	99	0	13

Note: MSE has to be multiplied by 0.001. Bias in percent of $\sigma = 1$, 0 stands for less than 1% $\sigma_j = 1, 1.2, 1.4, 1.6, 1.8$.

TABLE 5.4 Simulation Results for Model with Outliers

n_j	5				10				20			
	$\hat{\beta}_0$		$\hat{\beta}_1$		$\hat{\beta}_0$		$\hat{\beta}_1$		$\hat{\beta}_0$		$\hat{\beta}_1$	
	Bias	MSE	Bias	MSE	Bias	MSE	Bias	MSE	Bias	MSE	Bias	MSE
ILS	17	390	-6	45	17	205	-7	24	17	114	-7	14
IML	19	426	-6	47	19	225	-6	25	19	126	-7	14
RML	1	484	1	70	0	224	1	34	0	98	0	14

<u>Note</u>: MSE has to be multiplied by 0.001. Bias in percent of σ, 0 stands for less than 1% = 1. $\sigma = 1$ except for condition 5, $\sigma = 2$. $y_{rj} = 1.53 + x_j$.

or

$$\sigma_j = H(\beta' x_j, \theta) \tag{5.79}$$

where θ is an unknown constant.

Once an assumption such as (5.78) or (5.79) is made, one computes estimates of θ, then estimates

$$\hat{\sigma}_j^2 = H(x_j, \hat{\theta}) \quad \text{or} \quad \hat{\sigma}_j^2 = H(\underline{\beta}' x_j, \hat{\theta}) \tag{5.80}$$

and finally constructs a weighted estimate of $\underline{\beta}$. If H(•) is assumed to be completely unknown one can obtain information about H(•) by replications ($n_j > 1$). It was shown by Carroll (1982) that this method is more efficient than the ordinary least squares approach.

In the uncensored linear heteroscedastic model the ordinary least square estimator $\underline{\beta}_{LS}$ is unbiased and might therefore be applied in some practical cases despite the fact that more efficient estimators exist. In the censored linear heteroscedastic model the least square estimator $\underline{\beta}_{ILS}$ or the maximum likelihood estimator $\underline{\beta}_{IML}$, based on the assumption of homoscedasticity, are no longer unbiased.

In order to obtain an unbiased estimator or at least an asymptotically unbiased estimator for β, all estimation methods discussed in Chapter 5 can be modified to perform weighted estimators. This approach assumes, of course, that a model with replications is used where at least two uncensored observations at each design point x_j are available.

We shall consider a simple method to obtain unbiased estimators of $\underline{\beta}$ in the heteroscedastic case, where replications at each design point are available. This estimator consists of two steps. The first step is to obtain estimates of σ_j^2 and μ_j at each test condition. This can be done by using a BLU, ML, or RML-estimator. The conditional expectations of the censored observations is calculated next and substituted for the censored values. In the second step an ordinary least square estimator β_{LS} is calculated.

In a Monte Carlo simulation presented below we study the effect of heteroscedasticity on the ILS and the IML estimator. We also present the results for an estimator based on the procedure outlined above. This estimator uses the RML-estimator to obtain estimates of σ_j^2 and μ_j at each design point. We therefore denote this estimator as modified restricted maximum likelihood estimator, MRML. The results in Tables 5.3 and 5.4 are mixed. For small sample sizes

$n_j = 5$ the standard procedures IML and ILS result in smaller MSE than the modified MRML estimator, despite the large bias of the IML and ILS estimators. For increasing sample sizes the MRML estimator becomes clearly superior to the IML and ILS estimators. It should be noted that the effect of outliers on the estimates increases as the number of test conditions is reduced. The optimal design (see Section 5.6) which has only two test conditions is therefore not robust with respect to outliers and heteroscedasticity.

Robustness of the Censored Linear Model and Diagnostics

As we saw in the last section, the maximum likelihood estimator is not robust against heteroscedasticity. The same is true for Schmee and Hahn's (1979) least squares estimator. The development of robust estimators should therefore be the subject of future concern.

Tests and diagnostics are alternative methods of handling the problem of heteroscedasticity and outliers. Nelson (1981) considers a test for misspecification. Hall et al. (1982) discuss how diagnostic methods such as those introduced by Cook (1977) can be adapted to the censored linear model. In what follows we shall consider the special situation of accelerated life tests in which a few design points with replications are typical. Our approach is somewhat different from Cook's (1977). The main objective of accelerated life tests is to determine the median and pth percentile life time at an extrapolated point. We shall therefore study the effect of outliers on these estimates rather than their effect on $\hat{\underline{\beta}}$. Furthermore, we are dealing with a parametric model and thus we should use this information to detect outliers or heteroscedasticity.

The principle steps of the diagnostics suggested can be described as follows: We shall censor (Type II) possible outliers, rather than delete them, and study the effect on the prediction of median life and 10th percentile life at a particular point. This method is similar to Tiku's (1975) outlier test, and it is more efficient than deleting observations in a parametric model as shown by Schneider et al. (1985).

We shall now describe some effects of outliers and heteroscedasticity on estimation which can be explained from a theoretical point of view. Later we shall exemplify these results.

Effect of outliers. The effect of outliers in linear models was extensively studied in literature. In a censored linear model outliers become more consequential than in an uncensored linear model. In the latter case outliers influence the estimator $\hat{\underline{\beta}}$ directly. But the effect is not necessarily significant; it depends as well on the leverage of the outliers. In the censored case, however, this is different. Even when an outlier does not influence the estimator $\hat{\underline{\beta}}$ directly (because of its low leverage) this outlier might affect the estimator $\hat{\underline{\beta}}$ indirectly. Consider, for instance, an outlier in the middle of the design; it will hardly influence the slope in the uncensored case due to its low leverage. But assume now that some observations are censored at the end point of the design. To understand how this noninfluential outlier becomes more influential in the censored case, recall that the conditional expectations of censored observations enter the ML-estimator as observations. Now, even outliers with small leverage will increase the estimate $\hat{\sigma}^2$ of the residual variance substantially and hence increase the estimate $\hat{E}[y \mid y > y_r] = \hat{\mu} + \hat{\sigma}\tilde{W}(u_r)$ of the expectations of censored observations. Thus outliters affect the estimator $\hat{\underline{\beta}}$ indirectly, as they determine the magnitude of the "artificial" observations $\hat{E}[y \mid y > y_r]$. In the case of replications and heavy censoring the effect of outliers on $\hat{\underline{\beta}}$ might be massive.

Effect of heteroscedasticity. Heteroscedasticity in an uncensored linear model will have no effect on the bias of the MLE $\hat{\underline{\beta}}$. But in the censored linear model this is no longer true. The MLE will be heavily biased when heteroscedasticity is present. The reason for this is that the estimator for the expectations $\hat{E}[y \mid y > y_r] = \hat{\mu} + \hat{\sigma}\tilde{W}(u_r)$ becomes biased due to the use of the pooled estimate $\hat{\sigma}$. Hence, at design points with small σ_i^2 the estimates $\hat{E}$ will be too large and at design points with large σ_i^2 the estimates $\hat{E}$ will be too small.

Effect of total censoring. An analysis of the asymptotic variance reveals that the variance of the slope estimator is strongly dependent upon whether or not there are any uncensored observations at the extreme design points. The variance of the estimator $\hat{\underline{\beta}}$ depends, generally speaking, upon the method of censoring. An equal percentage of censored observations at all design points will primarily affect the bias and variance of the intercept rather than that of the slope. Unequal censoring at the design points, as is typical for accelerated life tests, will mainly increase the variance of the slope.

Table 5.5 gives the asymptotic variance of the IMLE of β for Type II censoring estimated from four test conditions ($x = 1, 2, 3, 4$; $\sigma = 1$) where the first three conditions are subject to equal censoring and the last condition is subject to 90% or 100% censorship. We see, for instance, that if 50% of the observations are censored at each design point, the asymptotic variances of the intercept and slope are $var(\beta_{0ML}) = 0.1907$ and $var(\beta_{1ML}) = 0.0244$. When, on the other hand, 10% of the observations at the first three conditions but all observations at the last condition are censored, the asymptotic variances increase to $var(\beta_{0ML}) = 0.2375$ and $var(\beta_{1ML}) = 0.0509$. Thus the loss of efficiency of β_{1ML} is more than 100%. We also notice that the variance of β_{1ML} is reduced significantly, namely to $var(\beta_{1ML}) = 0.0320$, by simply adding one uncensored observation at the last design point. Or, in other words, the variance of the slope estimator increases about 60% when the last remaining observations at the fourth design point is censored as well.

These results become very consequential for accelerated life tests since the observations in such tests tend to be increasingly censored with decreasing stress and the experimenter might tend to stop the experiment at the lowest stress before an item on test has failed.

Example 5.1

Crawford (1970) considered an example of accelerated life tests in order to demonstrate the effect of censoring for the estimation of median life time. In this example the electrical insulation of 40 motorettes were tested at four temperatures: 220°C, 190°C, 170°C, and 150°C. The objective was to draw conclusions about the median life time of the motorettes at the working temperature 130°C. To do so the Arrhenius law, which is a chemical rate equation, was applied. This law suggests that the mean logarithmic life as a function of the absolute temperature T_a can be written as

$$\text{mean log life} = \beta_0 + \beta_1/T_a \tag{5.77}$$

The observations of life time at the four temperatures were used to estimate β_0 and β_1. Then equation (5.77) was used to extrapolate in order to obtain an estimation of median life time at T = 130°C. To demonstrate the effect of censoring on estimation, data were collected at different times.

Table 5.6 gives the data which are available after 15, 16, 19, and 33 months. After 16 months, for instance, only 17 of the 40 observations are available; the remaining observations are censored; after 33 months only 3 observations are censored. One of the questions discussed by Crawford (1970) is how soon this test could have been discontinued without much loss of accuracy of the predicted median life time. His major conclusion is that "little additional information is gained by running the test beyond the point where the estimated common

TABLE 5.5 Asymptotic Variance of the MLE for Type II Censoring, 4 Test Conditions with 10 Replications

Percent of censored observations	Equal censoring at each design point		Equal censoring but last test condition has p% observations censored			
			p = 90%		p = 100%	
	var(β_0)	var(β_1)	var(β_0)	var(β_1)	var(β_0)	var(β_1)
10	1.527	0.203	1.840	0.320	2.375	0.509
20	1.573	0.209	1.873	0.327	2.664	0.550
30	1.644	0.217	1.924	0.335	2.554	0.543
40	1.748	0.229	2.001	0.346	2.712	0.572
50	1.907	0.244	2.123	0.361	2.950	0.611

Note: The entries have to be multiplied by 10^{-1}.

TABLE 5.6 Accelerated Life Test for 40 Motorettes

Temperature (°C)	Lifetime (hours)	Censoring times (hours)				
		15 months	16 months	19 months	21 months	33 months
220	408	336	528	700	700	17,661
220	408					
220	504					
220	504					
220	504					
220	600					
220	600					
220	648					
220	648					
220	696					
190	408	1,296	1,680	3,120	4,000	17,661
190	408					
190	1,344					
190	1,344					
190	1,440					
190	1,920					
190	2,256					

190	2,352					
190	2,596					
190	3,360					
170	1,764	5,112	5,448	6,792	7,632	17,661
170	2,772					
170	3,444					
170	3,542					
170	3,780					
170	4,680					
170	5,196					
170	6,206					
170	7,716					
170	7,884					
150	11,781	7,392	8,064	9,429	11,421	17,661
160	12,453					
150	13,897					
150	14,469					
150	15,891					
150	17,325					
150	17,325					
150						
150						
150						

Source: From Crawford, 1970.

standard deviation, the regression coefficients, and the estimated errors on the regression line are all essentially stabilized" and "all of these parameters reached this condition by the analysis at the 16th month or at least by the 17th month." He therefore concludes: "It appears that a relatively accurate estimate of the life of an insulation system can be drawn when approximately half of the units have failed."

The example considered by Crawford has attracted significant attention in literature. Nelson and Hahn (1972 and 1973), Schmee and Hahn (1979) and Aitkin (1981) considered this example to demonstrate the performance of estimation algorithms for censored data. But all of these authors based their study on the 16th month data following Crawford's arguments that these data contain enough information to obtain sufficient accurate estimates.

Figure 5.2 shows the logarithmic life time versus the reciprocal absolute temperature for the 33rd month data. The first impression is that the variances at 220°C and 150°C seem to be smaller than the variances at 170°C and 190°C. We therefore tested the 33rd month, ignoring the three censored observations at 150°C. The Levene (1960) test does not reject the hypothesis of homoscedasticity for $\alpha = 0.05$.

The two observations of 408h at 190°C are very likely to be outliers, i.e., the failure of those units is not caused by the temperature. In order to detect the influence of these observations we apply the diagnostic method introduced above, i.e., the lowest two observations at 190°C for the 15th, 16th, 19th, 21st, and 33rd month data are Type II censored. The results of a regression using the Arrhenius model (5.77) with log basis 10 are presented in Table 5.7 and Table 5.8.

As we see from Table 5.7 the residual variance is reduced to about 40% when the two observations at 190°C are Type II censored. Figure 5.3 shows the effect that the two outliers have on the estimated median life and 10th percentile life at working temperature 130°C.

First we notice that the influence of the two outliers on median life at 130°C is not significantly high for the 33rd month data. The median life at 130°C for the 33rd month data where only 3 observations are censored is 47,200 and 48,100 hours respectively. For the 16th month data the influence of the two outliers is fairly large. The predicted median life at 130°C is 46,700 hours when all the 16th month data are used and only 36,400 hours when the two observations at 190°C are censored. This drastic effect is due to the leverage, not only of the outliers, but of the 10 censored observations at 150°C. These censored observations are replaced by their estimated expectations which are 18,000 hours when all the 16th month observations are used and 14,000 hours when the two observations at 190°C are censored. Due to the high leverage of these 10 values, $\hat{\beta}_1$ is mainly determined by them.

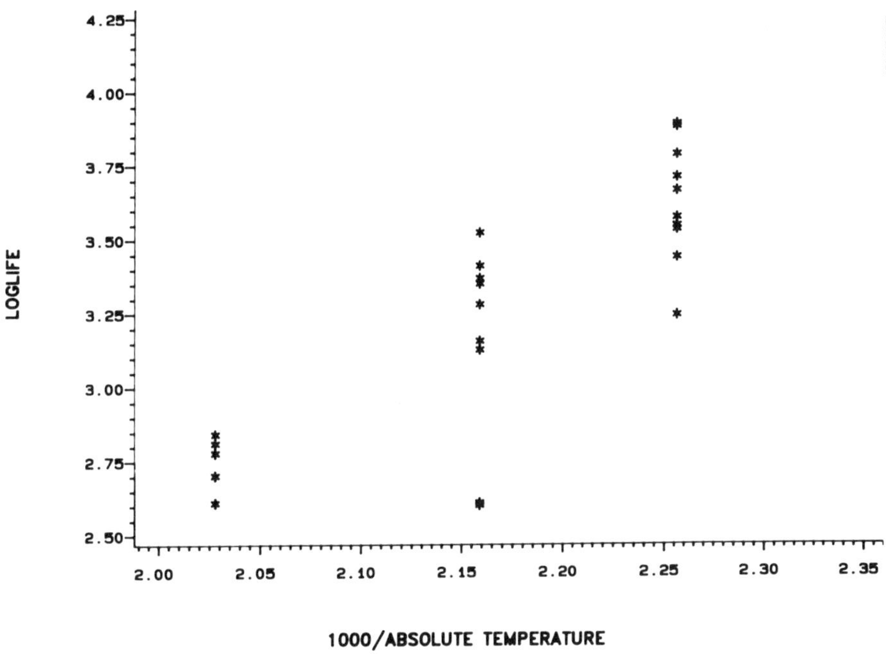

FIGURE 5.2 Motorettes data.

TABLE 5.7 MLE $\hat{\beta}$ for All Data and for Two Outliers at 190°C Censored

	$\hat{\beta}_0$		$\hat{\beta}_1$		$\hat{\sigma}$	
Month	W/o outl.	All data	W/o outl.	All data	W/o outl.	All data
15	-5.084	-4.887	3.883	3.853	0.2123	0.3980
16	-5.496	-6.010	4.055	4.306	0.1671	0.2757
19	-5.955	-6.541	4.273	4.539	0.1690	0.2359
21	-6.355	-6.934	4.462	4.725	0.1829	0.2405
33	-6.150	-6.379	4.364	4.460	0.1539	0.2128

Note: $\hat{\sigma}$ is adjusted by $\sqrt{n_0/(n_0 - 2)}$ to reduce bias.

TABLE 5.8 Median and 10th Percentile Life

	Median		10th Percentile	
Month	a	b	a	b
15	46.8	35.1	14.5	44.3
16	46.7	36.4	20.7	22.2
19	52.1	43.8	26.0	26.6
21	61.0	51.6	30.0	30.1
33	48.1	47.2	25.7	30.0

Note: a = all data estimates; b = two observations at 190°C Type II censored. Life is given in 1000 hours.

The confidence intervals for the median become necessarily smaller when the outliers are censored in view of the fact that $\hat{\sigma}$ also becomes smaller.

When considering the percentile life, the effect of outliers becomes more complicated. The pth percentile is given by antilog of $\beta_0 + \beta_1 x + u_p \hat{\sigma}$, where $x = 1000/(130 + 273.2)$. u_p is the p quantile of the standard normal distribution. For $p < 0.5$ the quantile u_p becomes negative and two effects of outliers compensate each other, namely the outliers increase $\beta_0 + \beta_1 x$ as well as $\hat{\sigma}$, such that the difference of $\beta_0 + \beta_1 x$ and $|u_p|\hat{\sigma}$ is not greatly affected. On the other hand, when the influence of the outliers on $\hat{\underline{\beta}}$ diminishes as in the 33rd month data, while the influence on $\hat{\sigma}$ remains the same, we notice a significant difference between the 10th percentiles. We obtain 30,000 hours without outliers and 25,700 hours with the outliers.

As we have seen, censoring is a useful method for detecting the influence of outliers. The two observations of 408 hours at 190°C are very influential. It seems paradoxical that these extremely low life times remarkably increase the predicted median life at 130°C. There is no doubt that their influence should be bounded. We therefore prefer the predicted median life where the two values of 408 hours are censored.

Unfortunately the median life based on the 16th month data (36,400 hours) (without the outliers) seems to be a very poor prediction of the

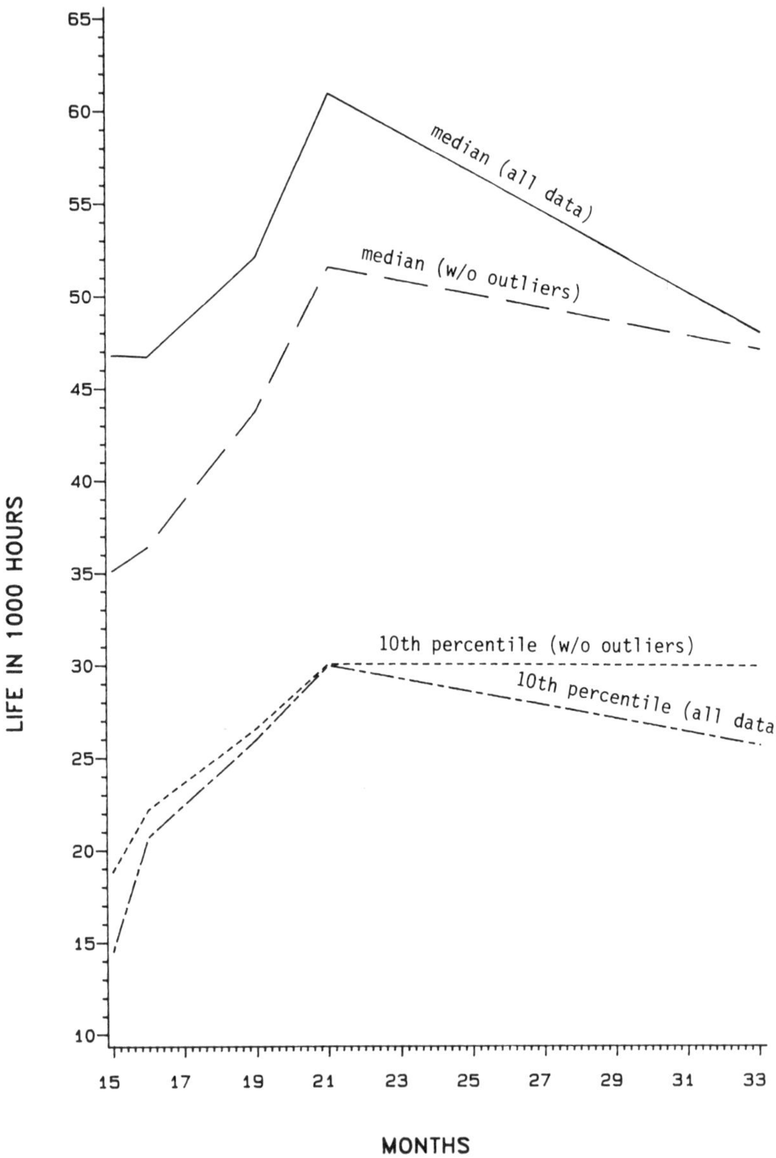

FIGURE 5.3 Median and 10th percentile for motorettes data.

33rd month median life, while the prediction, including the outliers, seems very accurate. But this is in accordance with our previous findings. The prediction based on the 16th and 19th months data are necessarily poor due to the total censoring at 150°C. As the asymptotic variance suggests, the estimates should improve when the first failure time at 150°C is added to the 16th month data. We thereby obtain a predicted median life at 130°C of 58,566 hours for the entire data estimate and 47,317 hours for the estimate where the outliers are censored. The estimates indeed only improve when the influence of the outliers is bounded.

Example 5.2

In order to study the behavior of the maximum likelihood estimators in the random censoring model a simulation study was performed by Schneider and Weissfeld (1986).

The linear model

$$Y_j = \underline{\beta}'\underline{x}_j + \epsilon_j \qquad j = 1, \ldots, n$$

was considered and it was assumed that the observed Y_j are generated by random censorship. Let the life times $Y_1, \ldots, Y_n$ be i.i.d. with distribution function F (not necessarily normal) and let the censoring times $C_1, \ldots, C_n$ be i.i.d. with distribution function G and further, under the assumption of random censorship, let $C_1, \ldots, C_n$ be independent of $Y_1, \ldots, Y_n$. We observe

$$Z_j = \min(Y_j, C_j)$$

and

$$\delta_j = \begin{cases} 1 & \text{if } Y_j \leq C_j \\ 0 & \text{if } Y_j > C_j \end{cases}$$

In the simulation study various beta distributions for F and G were assumed. It was found that the normal model tended to be fairly robust against departures from normality and in general performed quite well compared with the Buckley and James (1979) (BJ) nonparametric estimators. The IML of σ was in general less biased than the Buckley and James estimator for all types of assumed distributions for F and G.

TABLE 5.9 Buckley James and IML Estimates of the Slope and Intercept for the Stanford Heart Transplant Data

	$\hat{\beta}_0$	$\hat{\beta}_1$	$\hat{\sigma}$
IML	3.826	-0.02453	0.890
BJ	3.730	-0.02389	0.761

In order to demonstrate the relationship between the Buckley and James estimates and the maximum likelihood estimates based on the normal assumption for F, the Stanford heart transplant data (Miller and Halpern, 1982) were analyzed. Survival time ($\log_{10}$) as a function of age for survival times greater than 10 days was modeled. The results of the estimation are given in Table 5.9. The IML estimate of the slope $\underline{\beta}_1$ is almost identical with the BJ estimate, but the IML estimate of σ is significantly higher than the BJ estimate of σ. This result follows the pattern of the simulation study where the BJ estimator of σ always underestimates the standard deviation of Y.

5.6 OPTIMUM CENSORED ACCELERATED LIFE TESTS

In this section we shall briefly summarize the results on optimum plans for fitting a simple regression model to censored data. The literature on optimum test plans for least square fitting of linear regression models to complete data is summarized by Gaylor and Sweeny (1965). Nelson and Kielpinski (1976) give a brief summary of the literature on optimum plans for fitting regression models to censored data and they develop the theory for finding the optimal design for lognormal life distributions and Type I censoring.

Consider the model where the mean log-life is a linear function of (possibly transformed) stress x; that is

$$\mu(x) = \beta_0 + \beta_1 x \tag{5.81}$$

The log-life is assumed to be independently normally distributed with mean $\mu(x)$ and standard deviation σ. The problem is to find the design points which give the most precise estimate of the mean at a specified design stress μ_S. Nelson and Kielpinski (1972) provide an argument that the optimum plan uses just two stresses, x_L and x_H. They also state that the highest allowable test stress x_H must be specified; otherwise its optimum value would be infinitely large. We follow here the derivation of Nelson and Kielpinski (1976). The criterion which is used to find the optimal test stress x_L is that the large sample variance of the MLE of the mean at a specified design stress (x_S) be minimized. Since the optimal plan has only two test conditions, the parameters which have to be determined are the optimum low test stress

$$x_L^* = x_H + v^*(x_S - x_H) \tag{5.82}$$

and the optimum proportion p^* of test units to run at the optimum low test stress x_L. Nelson and Kielpinski (1972) provide a chart for p^*. Define the transformed stress $v = (x - x_H)/(x_S - x_H)$, where $v = 0$ for the highest stress, $v = 1$ for the design, and $v = v^*$ for the optimum stress. The Fisher information matrix for a sample of size n can then be written as the sum of two information matrices J_0 and J_v

$$n\underline{J} = (1 - p)n\underline{J}_0 + pn\underline{J}_v \tag{5.83}$$

where p is the proportion of the sample allocated to the low test stress v^*. The asymptotic covariance matrix of the ML estimators, which is the inverse of the Fisher information matrix, depends on the parameters p and v, the standardized censoring point at the highest stress $a = (y^c - \mu_H)/\sigma$, and the standardized difference of the means at design and highest stress $b = (\mu_S - \mu_H)/\sigma$. The variance of the ML estimator of the 100 pth percentile can now be expressed in terms of the asymptotic covariance matrix. In the reparameterized model the 100 pth percentile log life at the design stress x_S ($v = 1$) is

$$y_p(1) = \eta_0 + \eta_1 + u_p\sigma \tag{5.84}$$

where

$$\eta_0 = \beta_0 + \beta_1 x_H \tag{5.85}$$

and

$$\eta_1 = \beta_1(x_S - x_H) \tag{5.86}$$

Let B be the asymptotic covariance matrix of the estimators ($\hat{\eta}_0$, $\hat{\eta}_1$, $\hat{\sigma}$); then the variance of y_p is approximately

$$\text{var}[y_p(1)] = (1, 1, u_p)B(1, 1, u_p)' \tag{5.87}$$

The values of v and p that minimize the variance can be found numerically. A method is described by Nelson and Kielpinski (1972).

Unfortunately the minimal variance of the 100 pth percentile is a function of the unknown values a and b. To use the optimum plans one must provide approximate values of the parameter a and b. Values that are appreciably in error may result in a plan far from optimum. As pointed out by Nelson and Kielpinski (1976) the optimum plan with two test conditions is not robust against model misspecifications and errors in the data. Standard plans with three test conditions, as suggested by Nelson and Kielpinski (1976) and Meeker (1984), tend to be more robust. The results of Nelson and Kielpinski (1976) indicate, however, as a general rule, that more test units should be placed at low stress than at high stress.

REFERENCES

Aitkin, M. A. (1981). A note on the regression analysis of censored data, Technometrics 23, 161-163.

Ali, M. M. and Chan, L. K. (1964). On Gupta's estimates of the parameters of the normal distribution, Biometrika 51, 498-501.

Amemiya, T. (1973). Regression analysis when the dependent variable is truncated normal, Econometrica 41, 997-1016.

Amemiya, T. (1974). Multivariate regression and simultaneous equation models when the dependent variables are truncated normal, Econometrica 42, 999-1012.

Bement, T. R. and Williams, J. S. (1969). Variance of weighted regression estimators when sampling errors are independent and heteroscedastic, Journal of the American Statistical Association 65, 1369-1382.

Buckley, J. and James, I. (1979). Linear regression with censored data, Biometrika 69, 429-436.

Carroll, R. J. and Ruppert, D. (1982). Robust estimation in heteroscedastic linear models, Annals of Statistics 10, 429-441.

Cook, R. D. (1977). Detection of influential observations in linear regression, Technometrics 19, 15-18.

Crawford, D. E. (1970). Analysis of incomplete life test data on motorettes, Insulation/Circuits 16, 10, 43-48.

Dempster, A. P., Laird, N. M., and Rubin, D. B. (1977). Maximum likelihood from incomplete data via the EM algorithm (with discussion), Journal of Royal Statistical Society, Ser. B, 39, 1-38.

Fuller, W. A. and Rao, J. N. K. (1978). Estimation for a linear regression model with unknown diagonal covariance matrix, Annals of Statistics 6, 1149-1158.

Gaylor, D. W. and Sweeny, H. C. (1965). Design for optimal prediction in simple linear regression, Journal of the American Statistical Association 60, 205-216.

Glasser, M. (1965). Regression analysis with censored data, Biometrics 21, 300-307.

Harter, H. L. (1970). Order statistics and their use in testing and estimation, Vol. 2, Washington, U.S. Government Printing Office.

Hall, G. J., Rogers, W. H., and Pregibon, D. (1982). Outliers matter in survival analysis, Technical Report, University of Washington.

Hook, R. (1979). Discussion of: A simple method for regression analysis with censored data, Technometrics 21, 433-434.

Jacquez, J. A., Mather, F. J., and Crawford, C. R. (1968). Linear regression with non-constant, unknown error variances: Sampling experiments with least squares, weighted least squares, and maximum-likelihood estimators, Biometrics 24, 607-626.

Kalbfleisch, J. D. and Prentice, R. L. (1980). <u>The Statistical Analysis of Failure Time Data</u>, New York, Wiley.

Kendall, S. M. and Stuart, A. (1979). <u>The Advanced Theory of Statistics</u>, Vol. 2 (4th ed.), New York, Macmillan.

Lawless, J. F. (1983). Statistical methods in reliability (with discussion), Technometrics 25, 305-335.

Levene, H. (1960). <u>Robust Tests for Equality of Variances in Contributions to Probability and Statistics</u>, Ed. I, Orkin, Palo Alto, California, Stanford University Press, 278-292.

Miller, R. G. (1976). Least squares regression with censored data, Biometrika 63, 449-464.

Miller, R. G. and Halpern, J. (1982). Regression with censored data, Biometrika 69, 521-531.

Meeker, W. Q. (1984). A comparison of accelerated life test plans for Weibull and lognormal distributions and Type I censoring, Technometrics 26, 157-170.

Nelson, F. D. (1981). A test for misspecification in the censored normal model, Econometrica 49, 1315-1329.

Nelson, W. and Hahn, G. J. (1972). Linear estimation of a regression relationship from censored data, Part I — Simple methods and their application, Technometrics 14, 247-269.

Nelson, W. and Hahn, G. J. (1973). Linear estimation of a regression relationship from censored data, Part II — Best linear unbiased estimation and theory, Technometrics 15, 133-150.

Nelson, W. and Kielpinski, T. (1976). Theory for optimum accelerated life tests for normal and lognormal distributions, Technometrics 18, 105-114.

Nelson, W. and Kielpinski, T. (1972). Optimum accelerated life tests for normal and lognormal life distributions, General Electric Co. Corp. Research and Development TIS Report 72-CRD-215, IEEE Trans. on Reliability, R-24 (Dec. 1975).

Olsen, R. J. (1980). Approximating a truncated normal regression with the method of moments, Econometrica 48, 1099-1105.

Persson, T. and Rootzen, H. (1977). Simple and highly efficient estimators for a Type I censored normal sample, Biometrika 64, 123-128.

Prescott, P. (1979). Estimation of the standard deviation of a normal population from doubly censored samples using normal scores, Biometrika 57, 409-419.

Schmee, J. and Hahn, G. J. (1979). A simple method for regression analysis with censored data, Technometrics 21, 417-432.

Schmee, J. and Hahn, G. J. (1981a). A computer program for simple linear regression with censored data, Journal of Quality Technology 13, 264-269.

Schmee, J. and Hahn, G. J. (1981b). Correction to "A computer program for simple linear regression with censored data," Journal of Quality Technology 14, 235.

Schneider, H. and Weissfeld, L. (1986). Estimation in linear models with censored data, to appear in Biometrika.

Schneider, H., Weissfeld, L., and Carroll, R. J. (1985). A note on some simple influence diagnostics in accelerated life testing, Mimeo Series #1562, Department of Statistics at the University of North Carolina at Chapel Hill.

Scott, S. J. and Johnes, R. A. (1985). Quantifying seed germination responses to low temperatures: Variation among Lycopersicon SPP, Environmental and Experimental Botany 25, 1-9.

Tiku, M. L. (1975). A new statistic for testing suspected outliers, Communications in Statistics 4, 737-752.

Tiku, M. L. (1978). Linear regression model with censored observations, Communications in Statistics, Theory and Methods A7(13), 1219-1232.

Tobin, J. (1958). Estimation of relationships for limited dependent variables, Econometrica 26, 24-36.

Wolynetz, M. S. (1979). Maximum likelihood estimation in a linear model from confined and censored normal data, Algorithm AS139, Applied Statistics 28, 195-206.

CHAPTER 6

Inference with Censored Data

In this chapter we first discuss goodness-of-fit tests for censored samples. We then take up inferences concerning the parameters of the normal model, considering both interval estimation of the parameters and tests about them. Further test problems are discussed at the end of the chapter.

Section 6.1 deals with goodness-of-fit tests. The Kolmogorov-Smirnov test for goodness-of-fit was adapted to the censored case by Barr and Davidson (1973). Mihalko and Moore (1980) considered the Chi-square test of fit for Type II censored samples. Censored data versions of the Shapiro Wilk (1965) test were discussed by Smith and Bain (1976), Gerlach (1980), Shapiro and Brain (1981), and Verril and Johnson (1983). The Cramer-von Mises type statistics were modified by Pettitt and Stephens (1976) and Pettitt (1976) for application in the censored normal model.

Section 6.2 deals with inference concerning the parameters of the normal model. Censoring greatly complicates the distribution theory for the estimators, a situation which leads to asymptotic methods or Monte Carlo simulations. To overcome the mathematical difficulties Nelson and Schmee (1979) and Schmee et al. (1985) presented simulated confidence limits for selected sample sizes. Tiku (1978) discussed confidence intervals for symmetrically Type II censored samples. Schneider and Weissfeld (1986) gave asymptotically correct confidence limits. Alternate methods based on asymptotic theory are discussed by Kalbfleisch and Prentice (1980). The question of how large the sample size has to be in order to rely on asymptotic results cannot be answered in generality. It depends on the type of censoring, on the model and on the kind of test statistic which is used. The confidence limits suggested by Schneider and Weissfeld (1986), for instance, can be used for Type II censoring and sample sizes larger than 10. The likelihood ratio statistic, described below, does not give reliable results for sample sizes under 50, as simulation studies (Schneider and Weissfeld, 1986) show.

There are some other test problems which are not considered here. Tiku (1974b) discussed testing normality in multi-sample situations with censoring. In two other papers Tiku (1975, 1977) used Type II censoring to develop a statistic for testing suspected outliers. Testing group effects in analysis of variance models where samples are Type I or Type II censored, was investigated by Samford and Taylor (1959), Tiku (1973, 1982), and Tiku and Stewart (1977). Several papers, Tiku (1981, 1982) and Tiku and Singh (1982) deal with robust test statistics based on the modified maximum likelihood estimators.

6.1 GOODNESS-OF-FIT TESTS

A review of procedures for tests of distributional assumptions is given by Shapiro and Brain (1981). Several of these tests as well as the well-known Kolmogorov-Smirnov test and the chi-square test can be modified so that they can be used to test the goodness-of-fit with censored samples. The following sections give a review of these tests for testing normality.

Kolmogorov-Smirnov Test

The one sample Kolmogorov-Smirnov (K-S) test, based on the largest vertical difference between a hypothesized distribution function F and the observed sample distribution function F_n, has been considered by Barr and Davidson (1973) to test the goodness-of-fit with censored samples. If the standard Kolmogorov-Smirnov test were applied in the censored case the resulting test would be extremely conservative, since in the censored case the empirical distribution is not required to remain within the band beyond the censoring point. Barr and Davidson therefore suggest the following modifications of the K-S test statistic. Consider a singly censored sample at the right.

<u>Type I censoring</u>.

$$T_1 = \sup_{x \leq x_r} |\Phi(x) - F_n(x)| \tag{6.1}$$

Type II censoring.

$$T_2 = \sup_{x \leq x_{(r)}} |\Phi(x) - F_n(x)| \tag{6.2}$$

x_r is the truncation point in the uniform distribution while $x_{(r)}$ is the largest order statistic of a censored sample of size n from a uniform (0, 1) population. F_n is the corresponding sample distribution function.

Note that the distribution function has to be specified completely. If parameters of the normal distribution are estimated from the data used for the test the power of the test will change. The modified K-S test is invariant under the probability integral transformation of the underlying data and is thus, like the K-S test, distribution free. Tables for critical values were prepared by Barr and Davidson (1973).

Chi-square Test of Fit for Censored Data

When samples are Type I censored (multiply, doubly or singly) the chi-square test-of-fit applies in the usual way, since the censored observations fall in one or more fixed cells. This method is hence straightforward and not discussed here. The chi-square criterion is not generally applicable to testing the fit of Type II censored samples. Suppose that the (0, 1) interval is divided into m predetermined mutually exclusive intervals from which χ^2 is computed. Assume further only a right-censored sample of size n. If the largest, say kth, observation falls into any interval other than the mth, it is impossible to determine the cell index for the n - k censored values. For this case the χ^2 is theoretically not valid.

Mihalko and Moore (1980) extended the applicability of the chi-square test to Type II singly or doubly censored samples by employing sample percentiles as cell boundaries. We notice that this paper does not make distributional assumptions. In what follows we shall review the application to test the hypothesis that the population is normally distributed.

Consider a random sample of size n. We observe only the order statistics

$$x_{(n_\ell+1)} \leq x_{(n_\ell+2)} \leq \cdots \leq x_{(n-n_r)}$$

The smallest n_ℓ and the largest n_r values are censored. Let

$$0 = q_0 < q_1 < \cdots < q_{m-1} < q_m = 1$$

then m cells are defined as having boundaries

$$-\infty = \xi_{0n} < \xi_{1n} < \cdots < \xi_{m-1,n} < \xi_{mn} = \infty$$

where $\xi_{in} = X_{[(nq_i)]}$ is the sample q_i-quantile from $X_1, \ldots, X_n$. Set $q_1 = n_\ell/n$ when $n_\ell \geqslant 1$; otherwise $q_0 = n_\ell/n = 0$; similarly $q_{m-1} = (n - n_r)/n$ when $n_r \geqslant 1$; otherwise $q_m = (n - n_r)/n = 1$. The observed frequency N_{in} in the ith cell $E_i = (\xi_{i-1,n}, \xi_{in})$ is nonrandom, $N_{in} = (nq_i) - (nq_{i-1})$. According to the selection of q_1 and q_{m-1} the left-censored n_ℓ values and the right-censored n_r observations occupy the extreme cells.

Consider the composite-null hypothesis that the distribution of X_i is normal where the parameters μ and σ are estimated. The test procedure suggested by Mihalko and Moore requires that $\hat{\mu}$ and $\hat{\sigma}$ be estimators which are functions of the observed order statistics. Chi-square statistics for data-dependent cells are formed by ignoring the fact that the cells are functions of the data. Thus the ith estimated cell probability under H_0 is

$$\hat{p}_{in} = \Phi\left(\frac{\xi_{in} - \hat{\mu}}{\hat{\sigma}}\right) - \Phi\left(\frac{\xi_{i-1,n} - \hat{\mu}}{\hat{\sigma}}\right) \tag{6.3}$$

which is random, unlike the cell frequencies. Chi-square statistics are nonnegative definite quadratic forms in the standardized cell frequencies $(N_{in} - np_{in})/\sqrt{np_{in}}$.

Mihalko and Moore show that the vector of standardized cell probabilities is asymptotically multivariate normally distributed for quite a general class of estimators including the ML-estimators and the estimators of Chernoff, Gastwirth, and Johns (1967) discussed in Section 4.4. In what follows we assume a right censored sample of Type II. But the statistic can be easily extended to doubly censored samples.

We consider now only estimators of μ and σ which are asymptotically equivalent to the maximum likelihood estimators. Milhalko and Moore consider the original version of the Rao-Robson (1974) statistic. Let $u_i = (\xi_i - \hat{\mu})/\hat{\sigma}$, so that $p_i = \Phi(u_i) - \Phi(u_{i-1})$. Furthermore, set $\phi_i = \phi(u_i) - \phi(u_{i-1})$ and $\gamma_i = u_i\phi(u_i) - u_{i-1}\phi(u_{i-1})$ and define the matrix $\hat{\sigma}^{-2}D$, where the entries of D are

$$D_{11} = \Phi(u_{m-1}) - \gamma_m - \sum_{i=1}^{m-1} \phi_i^2/p_i \tag{6.4}$$

$$D_{12} = -(1 + u_{m-1}^2)\phi_m - \sum_{i=1}^{m-1} \phi_i \gamma_i/p_i \tag{6.5}$$

$$D_{22} = 2\Phi(u_{m-1}) - (1 + u_{m-1}^2)\gamma_m - \sum_{i=1}^{m-1} \gamma_i^2/p_i \tag{6.6}$$

Notice that for the sake of clarity we have omitted the subscript n. The Rao-Robson statistic is then

$$T_n = \sum_{i=1}^{m} (N_i - np_i)^2/np_i + [n\ \det(D)]^{-1}\left[D_{22}\left(\sum_{i=1}^{m} N_i\phi_i/p_i\right)^2 - 2D_{12}\left(\sum_{i=1}^{m} N_i\phi_i/p_i\right)\left(\sum_{i=1}^{m} N_i\gamma_i/p_i\right) + D_{11}\left(\sum_{i=1}^{m} N_i\gamma_i/p_i\right)^2\right] \tag{6.7}$$

The statistic T_n has a χ^2_{m-1} limiting null distribution. The proof is given by Mihalko and Moore (1980).

Shapiro-Wilk Test

The Shapiro-Wilk (1965) test is a regression-like test and many modifications were suggested since this test was introduced. A well-known but subjective method of testing for normality is the probability plot. A probability plot can be considered as the graphical representation of the regression of the ordered observations on the expected values of the order statistics from the standardized distribution. It can also be used for censored samples.

Consider a censored sample where $x_{(n_\ell+1)} \leqslant \cdots \leqslant x_{(n-n_r)}$ is the order statistic of the observed values of the sample of size n. Then under the null hypothesis the linear relationship

$$x_{(i)} = \mu + \sigma z_{i,n} + \epsilon \tag{6.8}$$

holds, where μ and σ are the mean and standard deviation of the normal population, respectively, and $z_{i,n}$ is the expected value of the ith order statistic from a standard normal distribution while ϵ is a random error. Thus departures from linearity indicate lack of fit. The probability plot often used in practice, however, is subjective and there is no Type I error associated with the procedure. Shapiro and Wilk (1965) suggested testing for linearity by comparing two estimates of σ, i.e., s^2 and the squared slope of an estimated regression line. Let us assume that we have fitted the regression line $\hat{x}_{(i)} = \hat{a} + \hat{b}u_{i,n}$ to the order statistics, then if the model is correct $\hat{b}^2$ provides an estimate of σ^2. If the null hypothesis is not true the linear model is not appropriate and thus $\hat{b}^2$ will be different from s^2. The ratio of both estimates therefore provides an appropriate test statistic which is independent of location and scale parameters and thus does not require knowing their values. But this statistic is not distribution-free.

Censored data versions of the Shapiro-Wilk test of normality were considered by Smith and Bain (1976), Gerlach (1980), Shapiro and Brain (1976), and Verril and Johnson (1983). Filliben (1975) and Ryan and Joiner (1973) noted that the Shapiro-Francia statistic could be written as the square of the correlation coefficient ρ

$$\rho^2 = \frac{\sum_{i=1}^{n} [(x_{(i)} - \bar{x})(z_{i,n} - \bar{x})]}{\left[\sum_{i=1}^{n} (x_i - \bar{x})^2 \sum_{i=1}^{n} (z_{i,n} - \bar{z})^2\right]^{1/2}} \tag{6.9}$$

associated with a linear normal model in which the sample order statistics, $x_{(i)}$, fulfill the relationship

$$x_{(i)} = a + bz_{i,n} + \epsilon \tag{6.10}$$

where ϵ is normally distributed with expectation zero. This interpretation of the Shapiro and Wilk test leads to alternative tests, where the expectations $z_{i,n}$ are substituted by the median of the ith order statistic from a sample of n standard normal random variables (Filliben, 1975), or by the inverse normal distribution $\Phi^{-1}[i/(n+1)]$ (Ryan and Joiner, 1973).

Gerlach (1980) considers the censored data versions of (6.9) which, for a doubly Type II censored sample, can be written as

$$\rho^2(n, n_r, n_\ell) = \frac{\sum_{i=n_\ell+1}^{n-n_r} (x_{(i)} - \bar{x})(z_{i,n} - \bar{z})}{\left[\sum_{i=n_\ell+1}^{n-n_r} (x_{(i)} - \bar{x})^2 \sum_{i=n_\ell+1}^{n-n_r} (z_{i,n} - \bar{z})^2\right]^{1/2}} \qquad (6.11)$$

where $n_0 = n - n_r - n_\ell > 1$ is assumed.

The hypothesis $H_0: F = \Phi$ has to be rejected if $\rho^2(n, n_r, n_\ell) < c(n, n_r, n_\ell, \alpha)$ where $P(\rho^2 < c \mid H_0) = \alpha$. Critical values $c(n, n_r, n_\ell, \alpha)$ are given by Gerlach (1980) who used a Monte Carlo simulation of 1000 samples to determine selected values $c(n, n_r, n_\ell, \alpha)$ for $10 \leq n \leq 100$ and $\alpha = 0.01, 0.05, 0.1, 0.25, 0.5$. Verrill and Johnson (1983) considered the version where the expectations $z_{i,n}$ are substituted by the inverse normal cdf $\Phi^{-1}[i/(n+1)]$. Critical values for this test are given by Verrill (1981). These tables are also generated by Monte Carlo simulations. The asymptotic distribution of this modified test statistic is given by Verrill and Johnson (1983). Monte Carlo studies by these authors suggest that the large sample approximations are fairly good when $n_0 = n - n_r - n_\ell > 50$.

Cramer-von Mises Statistics

The Cramer-von Mises type statistics W_n^2 (Cramer-von Mises) and A_n^2 (Anderson-Darling) and the Watson statistic U_n^2 were modified by Pettitt and Stephens (1976) and Pettitt (1976) so that tests of goodness-of-fit could be made for the simple hypothesis with censored data.

Consider a random sample $X_1, \ldots, X_n$ taken from a population with cdf $\Phi[(x - \mu)/\sigma]$. We note that the Cramer-von Mises test statistics do not depend on the specific form of the distribution; hence the normal distribution might be substituted by any other continuous distribution function $G(x)$. The Cramer-von Mises test is based on $t_i = G(x_i)$ instead of on original data. Suppose we want to test the hypothesis $H_0: G(x) = G_0(x)$, where $G_0(x)$ is completely specified as in the case of $\Phi[(x - \mu)/\sigma]$. The null hypothesis is equivalent to testing that the observations $t_i = G_0(x_i)$ $(i = 1, \ldots, n)$ come from the uniform distribution on $(0, 1)$. Let $F(t)$ be the distribution function of $T = G_0(X)$ and $F_n(t)$ the empirical distribution of T, determined by $G_0(x_1), \ldots, G_0(x_n)$. The Cramer-von Mises type statistics is then

$$\omega_n^2 = n \int_0^1 [F_n(t) - t]^2 \psi(t)\, dt \tag{6.12}$$

If $\psi(t) = 1$, ω_n^2 is the Cramer-von Mises statistic W_n^2 and if $\psi(t) = [t(1 - t)]^{-1}$, ω_n^2 is the Anderson-Darling statistic A_n^2. The Watson statistic U_n^2 is given by

$$U_n^2 = n \int_0^1 \left\{F_n(t) - t - \int_0^1 [F_n(s) - s]\, ds\right\}^2 dt \tag{6.13}$$

For censored samples, observations are only available in certain intervals. For Type I censoring $\{G_0(x) = \Phi[(x - \mu_0)/\sigma_0]\}$ define $p_\ell = \Phi[(x_\ell - \mu_0)/\sigma_0]$ and $p_r = \Phi[(x_r - \mu_0)/\sigma_0]$ while for Type II censoring set $x_\ell = x_{(n_\ell+1)}$ and $x_r = x_{(n-n_r)}$. The modifications of the Cramer-von Mises type statistics considered by Pettitt and Stephens (1976) are

$$W_n^2(p_\ell, p_r) = n \int_{p_\ell}^{p_r} [F_n(t) - t]^2\, dt \tag{6.14}$$

$$A_n^2(p_\ell, p_r) = n \int_{p_\ell}^{p_r} [F_n(t) - t]^2/[t(1 - t)]\, dt \tag{6.15}$$

For the Watson statistic only singly censored samples were considered.

$$U_n^2(p_r) = n \int_0^{p_r} \left\{F_n(t) - t - 1/p_r \int_0^{p_r} [F_n(s) - s]\, ds\right\}^2 dt \tag{6.16}$$

Pettitt and Stephens (1976) derived the asymptotic distribution of these statistics when the distribution function $G_0(x)$ is completely specified. Pettitt (1976) considers the asymptotic distribution when some parameters of the cdf $G_0(x)$ have to be estimated.

Monte Carlo experiments (Pettitt, 1976) for small sample sizes confirm that the distributions of $W_n^2(p)$ and $A_n^2(p)$ (i.e., for singly censored samples of size n) converge quickly to their asymptotic distributions when censoring is of Type I. But the convergence is slow when the samples are Type II censored.

Pettitt and Stephens (1976) give tables of asymptotic percentage points for the completely specified case while Pettitt (1976) gives tables for the case where the parameters of the normal distribution have to be estimated. The percentage points given by Pettitt (1976) vary with the estimators applied to estimate μ and σ.

6.2 TESTS AND INTERVAL ESTIMATION

In this section we take up inferences concerning the parameters of a normal model. Throughout the section we use $\theta = (\theta_1, \ldots, \theta_k)$ as a parameter vector where $\theta_1 = \mu$ and $\theta_2 = \sigma$ in the simple normal model while $\theta_1 = \beta_0$, $\theta_2 = \beta_1, \ldots, \theta_{k-1} = \beta_{k-2}$ and $\theta_k = \sigma$ for the regression model.

In the following subsection we consider the main asymptotic results that apply to the likelihood function and the maximum likelihood estimators. We then discuss some tests based on these results; finally, we study inferences for small samples. No derivations are given here; the reader is referred to Serfling (1980) for a more complete discussion.

Large Sample Theory

Let us assume that the sample consists of n independent observations taken from a normal population, $n - n_0$ of which are censored. Let $\theta = (\theta_1, \ldots, \theta_k)$ be the parameter vector and define the likelihood function of θ as

$$L_n(\theta) = \prod_{i=1}^{n} f(x_i; \theta) \tag{6.17}$$

where $L(x_i; \theta)$ is the pdf of x_i. For the purpose of this subsection we shall use the same notation for censored and uncensored observations in the likelihood function.

We define the vectors

$$a_{n\theta} = \left[\frac{1}{n}\sum_{i=1}^{n} \partial \log f(X_i;\theta)/\partial\theta_1, \ldots, \frac{1}{n}\sum_{i=1}^{n} \partial \log f(X_i;\theta)/\partial\theta_k\right] \tag{6.18}$$

and

$$d_{n\theta} = \hat{\theta}_n - \theta = (\hat{\theta}_{n1} - \theta_1, \ldots, \hat{\theta}_{nk} - \theta_k) \tag{6.19}$$

where $\hat{\theta}_n = (\hat{\theta}_{n1}, \ldots, \hat{\theta}_{nk})$ denotes a consistent, asymptotically normal, and asymptotically efficient sequence of solutions to the likelihood equations. Furthermore, let J_θ be the Fisher information matrix. Then the following lemma will hold:

Lemma A. Let $X_1, X_2, \ldots$ be i.i.d. with distribution F . Then (under appropriate regularity conditions, which are fulfilled in the Type I and Type II censored normal models)

1. $n^{1/2} a_{n\theta} \xrightarrow{d} N(0, J_\theta)$
2. $n^{1/2} d_{n\theta} \xrightarrow{d} N(0, J_\theta^{-1})$
3. $n a_{n\theta} J_\theta^{-1} a'_{n\theta} \xrightarrow{d} \chi_k^2$
4. $n d_{n\theta} J_\theta d'_{n\theta} \xrightarrow{d} \chi_k^2$

Proof. (1) follows directly from the multivariate Lindberg-Levy central limit theorem; (2) follows from the theorem (Serfling, 1980, p. 145), namely that under regularity conditions maximum likelihood estimates are strongly consistent, asymptotically normal and asymptotically efficient; (3) and (4) follow from (1) and (2), respectively.

Lemma B. Let $X_1, X_2, \ldots$ be i.i.d with distribution F . Then (under appropriate regularity conditions)

$$2[L_n(\hat{\theta}_n) - L_n(\theta)] \xrightarrow{d} \chi_k^2$$

Proof. Lemma B is a direct consequence of Lemma A (4) and

$$L_n(\hat{\theta}_n) - L_n^n(\theta) - \frac{1}{n} nd_{n\theta} J_\theta d'_{n\theta} \xrightarrow{p} 0$$

The latter can be shown by Taylor expansion (see Serfling, 1980, p. 154).

Test Statistics for a Single Null Hypothesis

We consider several tests which follow immediately from Lemma A and Lemma B.

Tests based on the MLE. The statistic

$$T_n = n^{1/2} d_{n\theta} \tag{6.20}$$

is asymptotically normally distributed [Lemma A (2)] with expectation zero and covariance matrix J_θ^{-1}. Therefore for large n, an elliptical confidence region for θ can be formed by using the limit distribution $N(0, J_\theta^{-1})$.

The statistic

$$W_n = nd_{n\theta_0} J_{\theta n} d'_{n\theta_0} \tag{6.21}$$

which was introduced by Wald (1943) is asymptotically χ_k^2 distributed. This follows from Lemma A (4) and the fact that $J_\theta \xrightarrow{wp1} J_\theta$.

Tests based on the score vector. The statistic

$$S_n = n^{1/2} a_{n\theta} \tag{6.22}$$

is asymptotically normal with mean zero and covariance matrix J_θ [Lemma A (1)]. Elliptical confidence regions can therefore be formed using the limit distribution.

The statistic

$$V_n = na_{n\theta_0} J_\theta^{-1} a_{n\theta_0} \tag{6.23}$$

was introduced by Rao (1947) and has an asymptotic χ_k^2 distribution, which follows from Lemma A (3).

Likelihood ratio test. The likelihood ratio statistic

$$\Lambda_n = \frac{L_n(\theta_0)}{L_n(\hat{\theta})} \tag{6.24}$$

where $\hat{\theta}$ is the MLE of θ was introduced by Neyman and Pearson (1928). For asymptotic considerations the statistic

$$\lambda_n = -2 \log \Lambda_n \tag{6.25}$$

turns out to be more convenient. From Lemma B we ascertain that λ_n is asymptotically χ_k^2 distributed.

Test Statistics for a Composite Null Hypothesis

Consider the vector of parameters $\theta = (\theta_r, \theta_s)$, where it is to be understood that θ_r and θ_s may each consist of more than one parameter. The hypothesis to be tested is

$$H_0 : \theta_r = \theta_{r0}$$

the parameter θ_s being left unspecified by the hypothesis tested. We shall only consider the likelihood ratio statistic. The other statistics described above can be adapted to test for a composite null hypothesis in a similar fashion (see Serfling, 1980, pp. 156-160).

Let $\hat{\theta}_s(\theta_{r0})$ be the MLE of θ_s given $\theta_r = \theta_{r0}$; furthermore $\hat{\theta}$ denotes the MLE of $\theta = (\theta_r, \theta_s)$; then

$$\Lambda_n = L_n[\theta_r, \hat{\theta}_s(\theta_{r0})]/L_n(\hat{\theta}) \tag{6.26}$$

The statistic

$$\lambda_n = -2 \log \Lambda_n \tag{6.27}$$

is asymptotically χ_r^2 distributed with degrees of freedom given by the dimension of θ_r under the hypothesis $\theta_r = \theta_{r0}$.

All the above mentioned statistics can be used to obtain confidence regions for μ or σ. But those regions are rather cumbersome to compute since the maximum likelihood estimators are not available in closed form. Even more serious is that the small sample behavior of the statistics for censored data is hardly known. It will depend on the degree of censoring. For instance, simulation studies by Schneider and Weissfeld (1986) reveal that the likelihood ratio statistic gives poor results for sample sizes up to 50 even if censoring is moderate. Before applying asymptotic tests or confidence intervals, finite sample behavior should be studied. In the following subsection we consider simplified test statistics which are applicable for small sample sizes.

Confidence Intervals for μ

Throughout this subsection we assume that X is normally distributed with unknown mean μ and unknown variance σ^2. Suppose a sample of size n is given where the n_ℓ smallest and n_r largest observations are censored. We restrict the discussion to Type II censoring. The maximum likelihood estimators μ_{ML} and σ_{ML} have asymptotic variance

$$\mathrm{var}(\mu_{ML}) = \frac{\sigma^2}{n_0} J_{22}/(J_{11}J_{22} - J_{12}^2) \tag{6.28}$$

and

$$\mathrm{var}(\sigma_{ML}) = \frac{\sigma^2}{n_0} J_{11}/(J_{11}J_{22} - J_{12}^2) \tag{6.29}$$

where $n_0 = n - n_r - n_\ell$ and J_{ij} are the elements of the Fisher information matrix.

Elliptical confidence regions for (μ_{ML}, σ_{ML}) can be constructed by using the asymptotic results presented above. For small sample sizes, degree of freedom adjustments have to be made when using the quantities of the student distribution. To obtain simplified joint confidence intervals the Bonferrori inequality might be used. But if μ_{ML} and σ_{ML} have a large covariance the Type I error tends to be too small.

In the case of symmetrical censoring μ_{ML} and σ_{ML} are uncorrelated and confidence intervals and tests can be easily constructed.

We shall describe two methods for obtaining approximate confidence intervals for μ when σ is unknown. The first method is due to

Tiku (1971), who uses the t-distribution with n_0 - 1 degrees of freedom to approximate the distribution of a normalized ratio of $\hat{\mu}$ and $\hat{\sigma}$. This method is only applicable for symmetrical censored samples. The second approach is based on the asymptotic distribution of $\hat{\mu} + k\hat{\sigma}$.

We shall now examine these two methods in greater detail. We first notice that bias-corrected estimates $\hat{\sigma}$ and $\hat{\mu}$ give more accurate confidence intervals for μ than the original MLE. We therefore apply the bias-correction methods discussed in Section 4.5. We also notice that the bias-corrected MML as well as the bias-corrected RML give about the same Type I error as the bias-corrected MLE.

The t-Statistic for Symmetrical Type II Censoring

Let $\hat{\sigma}_a$ be a bias corrected estimator of $\hat{\sigma}$, which is consistent, asymptotically efficient, and normally distributed. Then

$$T_{\sigma_a} = (n_0 - 1)\hat{\sigma}_a^2/\sigma^2 \tag{6.30}$$

is asymptotically chi-square distributed with n_0 - 1 degrees of freedom (Tiku, 1978). Let μ_{ML} be the MLE of μ; then the statistic

$$T_t = \frac{(\mu_{ML} - \mu)/\sigma \sqrt{n_0/\gamma}}{\sqrt{\hat{\sigma}_a^2/\sigma^2}} = \frac{\mu_{ML} - \mu}{\hat{\sigma}_a} \sqrt{n_0/\gamma} \tag{6.31}$$

is approximately student-distributed with $f = n_0 - 1$ degrees of freedom, where

$$\gamma = J_{22}/(J_{11}J_{22} - J_{12}^2) \tag{6.32}$$

We notice that μ_{ML} can be replaced with any estimator of μ which is asymptotically equivalent to the MLE.

We might, for instance, use Tiku's (1967) MML. We also could consider the RML estimator of μ, which gives similar results for the confidence intervals as simulation results show. Confidence intervals and tests for μ and σ can now be constructed in the usual way.

To study the effect of unsymmetrical censoring on the probability of the Type I error for a t-test and on the confidence level of a confidence interval, a Monte Carlo simulation was performed by Schneider and

Weissfeld (1985). They show that the t-statistic gives only reliable results for symmetrical censoring. For unsymmetrical censoring the t-statistic becomes skewed to the side of heavier censoring. The results of the simulation study also suggest that the bias correction (4.179), which is based on Saw's (1961) table, performs better than the degree of freedom adjustment (4.180) for $\hat{\sigma}$.

The z-Statistic for General Type II Censoring

For non-symmetrical censoring the estimators $\hat{\mu}$ and $\hat{\sigma}$ are correlated and the confidence intervals have to be unsymmetrical to obtain equal probabilities $\alpha/2$ in both tails. We consider the random variable $T_z = \hat{\mu} \pm \hat{\sigma} z$ and determine z_1 and z_2 such that for $\mu = \mu_0$

$$P(\hat{\mu} + z_1\hat{\sigma} < \mu) = \alpha/2 \tag{6.33}$$

and

$$P(\mu - z_2\hat{\sigma} > \mu) = \alpha/2 \tag{6.34}$$

Hence, the probability that the interval $(\hat{\mu} - z_2\hat{\sigma},\ \hat{\mu} + z_1\hat{\sigma})$ includes μ is $1 - \alpha$. For appropriate estimators $\hat{\mu}$ and $\hat{\sigma}$ (ML, RML, MML) the statistic T_z is asymptotically normal with mean

$$E[T_z] = \mu \pm z\sigma \tag{6.35}$$

and variance

$$\operatorname{var}[T_z] = \operatorname{var}(\hat{\mu}) + z^2 \operatorname{var}(\hat{\sigma}) \pm 2z \operatorname{cov}(\hat{\mu}, \hat{\sigma})$$

The random variable

$$Z = \frac{T_z - (\mu \pm z\sigma)}{\sqrt{\operatorname{var}(T_z)}} \tag{6.36}$$

is therefore asymptotically standard normally distributed.

Let $\operatorname{var}(\hat{\mu}) = (\sigma^2/n)\gamma_{11}$, $\operatorname{var}(\hat{\sigma}) = (\sigma^2/n)\gamma_{22}$ and $\operatorname{cov}(\hat{\mu}, \sigma) = (\sigma^2/n)\gamma_{12}$; then z_1 and z_2 are solutions to the equations

TABLE 6.1 Simulated Percentages for $\alpha = 0.10$

		10		20	
n		One-sided	Two-sided	One-sided	Two-sided
p_ℓ	p_r	$\alpha/2$	α	$\alpha/2$	α
0	1	5.4	10.7	5.1	11.2
	2	5.1	10.7	4.7	10.1
	3	5.3	10.5	5.3	10.7
	4	5.3	11.3	5.0	10.3
	5	4.9	10.9	5.0	10.4
	6	5.6	12.1	5.2	10.3
	7	5.1	11.4	5.7	11.5
	8	3.8*	6.2*	4.8	10.5
1	1	4.9	10.4	5.6	10.0
	2	5.1	10.6	4.6	10.4
	3	5.1	10.6	4.8	10.2
	4	5.4	11.4	4.5	10.4
	5	5.4	11.3	4.5	9.8
	6	5.0	10.4	5.0	10.6
	7	5.9*	10.1*	5.4	10.4
2	2	5.6	11.2	5.4	10.2
	3	6.2	11.9	5.3	10.5
	4	5.4	12.1	4.8	10.9
	5	6.2	11.4	5.2	10.9
		8.7*	14.6*	5.4	10.4
3	3	6.1	12.6	5.5	10.9
	4	7.1	13.1	5.7	11.7
	5	11.4*	18.3*	6.3	11.8

Note: The results are based on 10,000 samples; (*) marks cases where the denominators of (6.39) and (6.40) become negative; the one-sided percentages are above the 95th percentile; two-sided percentages are below the 5th and above the 95th percentile.

$$\Phi\left\{ -z_1[n(\gamma_{11} + z_1^2\gamma_{22} + 2z_1\gamma_{12})]^{1/2} \right\} = \alpha/2 \tag{6.37}$$

and

$$\Phi\left\{-z_2[n/(\gamma_{11} + z_2^2\gamma_{22} - 2z_2\gamma_{12})]^{1/2}\right\} = \alpha/2 \tag{6.38}$$

Define

$$A = -\gamma_{12}u^2_{\alpha/2}/(n - u^2_{\alpha/2}\gamma_{22}) \tag{6.39}$$

and

$$B = \gamma_{11}u^2_{\alpha/2}/(n - u^2_{\alpha/2}\gamma_{22}) \tag{6.40}$$

Then we obtain

$$z_1 = -A + \sqrt{A^2 + B} \tag{6.41}$$

$$z_2 = A + \sqrt{A^2 + B} \tag{6.42}$$

Table 6.1 gives some simulation results for the proposed statistic T_z when bias corrected MLEs are used. We see that the statistic controls the probabilities in both tails. The performance for small sample sizes is remarkably good and even for symmetrical censoring (Schneider and Weissfeld, 1986) the z-statistic turns out to be not inferior to the t-statistic. In some cases, marked by an asterisk when the sample size is small the denominator in (6.39) and (6.40) becomes negative and the confidence limits deterioriate. We note that for $n \longrightarrow \infty$ the Z- and the t-statistic give the same result. However, T_z converges much faster to normality than T_t.

Confidence Intervals for σ

Several confidence limits for the standard deviation were suggested in literature. Tiku (1978), considering symmetrically censored samples, uses the chi-square distribution with $n_0 - 1$ degrees of freedom to approximate the distribution of the variance estimator $\hat{\sigma}^2$, where n_0 is the number of uncensored observations.

The standard large sample confidence interval for σ, which uses the asymptotic normality of $\hat{\sigma}$, is not to be recommended for small sample sizes. Schmee et al. (1985) consider $\log(\hat{\sigma})$, which has a less skewed distribution for finite sample sizes. Let $\hat{\sigma}$ be an asymptotic efficient and normally distributed estimator of σ, with asymptotic variance $\sigma^2\gamma_{22}/n$. Then $\log(\hat{\sigma})$ is asymptotically normally distributed with variance γ_{22}/n (Serfling, 1980, p. 118). Large sample confidence limits for $\log(\sigma)$ are given by

$$\log(\sigma_L) = \log(\hat{\sigma}) + u_{\alpha/2}\sqrt{(\gamma_{22}/n)} \tag{6.43}$$

$$\log(\sigma_U) = \log(\hat{\sigma}) - u_{\alpha/2}\sqrt{(\gamma_{22}/n)} \tag{6.44}$$

where $u_{\alpha/2}$ is the $(100\alpha/2)$th standard normal percentile. Notice that the length of these confidence intervals for $\log(\sigma)$ are independent of the estimate $\hat{\sigma}$, given Type II censoring. Schmee et al. (1985) transform (6.43) and (6.44) to obtain confidence limits for σ

$$\sigma_L = \hat{\sigma}\exp\left[u_{\alpha/2}\sqrt{(\gamma_{22}/n)}\right] \tag{6.45}$$

$$\sigma_U = \hat{\sigma}/\exp\left[u_{\alpha/2}\sqrt{(\gamma_{22}/n)}\right] \tag{6.46}$$

These confidence limits are usually more accurate than the limits obtained by standard asymptotic arguments, i.e., $\hat{\sigma}\left[1 \pm u_{\alpha/2}\sqrt{(\gamma_{22}/n)}\right]$

For linear unbiased estimators, Mann, Schafer, and Singpurwalla (1974) give approximate confidence limits for σ. Their methodology can be roughly described as follows. Under certain regularity conditions for location-scale parameter distributions there are v_i that can be used to divide the difference of the (i+1)st and ith ordered observation and to convert these differences to approximate exponential variates with mean and variance approximately 1. A result of Patnaik (1949) then guarantees that a linear combination of these variates has approximately a weighted chi-square distribution with degrees of freedom given by $2/C$, where $\text{var}(\hat{\sigma}) = \sigma^2 C$. Since the number of degrees of freedom would not be expected to be an integer, the Wilson-Hilferty (1931) approximation (see Johnson and Kotz, 1970, p. 176) of chi-square to normality can be used to obtain approximate confidence limits. Using this approach, Mann, Schafer, and Singpurwalla (1974) derive the following confidence limits:

$$\sigma_L = \hat{\sigma}/\left[1 - C/9 - u_{\alpha/2}\sqrt{(C/9)}\right]^3 \tag{6.47}$$

$$\sigma_U = \hat{\sigma}/\left[1 - C/9 + u_{\alpha/2}\sqrt{(C/9)}\right]^3 \tag{6.48}$$

These confidence limits originally derived for linear unbiased estimators can also be applied for the bias corrected MLE, since, as was shown by Plackett (1958), the MLE $\hat{\sigma}$ is asymptotically linear. Notice that for the MLE we obtain $C = \gamma_{22}/n$, where γ_{22} is the variance factor tabulated in Table A3 of Appendix A.

In a simulation study for sample sizes 10, 20, and 50 we compared the suggested confidence limits using the MLE. The following excerpt of the simulation results (Table 6.2) gives an impression of the small sample behavior of these limits.

It seems that the confidence limits of Mann, Schaffer, and Singpurwalla (1974) give more accurate results than the other methods for the MLE. Tiku's chi-square approximation does not work satisfactorily for singly censored samples, and the limits based on the asymptotic normality of $\log(\hat{\sigma})$ converge very slowly to the correct confidence level. We also notice that the sample size has to be at least 20 in order to obtain reliable results for moderate censoring. For heavy censoring the sample size has to be even higher.

TABLE 6.2 Simulated Percentages Below the Lower and Above the Upper Confidence Limits ($\alpha/2 = 0.05$)

		Chi-square		Weighted chi-square		Log-transformed	
n	n_r/n	Below	Above	Below	Above	Below	Above
10	0.3	9.1	6.3	5.9	9.1	3.5	12.8
	0.5	8.0	7.0	5.6	11.3	2.5	15.1
20	0.3	6.8	6.1	4.9	6.7	3.6	9.5
	0.5	9.5	6.2	7.0	7.4	5.2	10.5
50	0.3	7.1	5.1	4.8	5.3	3.6	6.5
	0.5	9.8	6.7	6.3	6.6	4.4	7.7

Confidence Limits for μ and σ Obtained by Monte Carlo Method

For small samples confidence intervals based on asymptotic theory are usually too small. Monte Carlo methods can be used to obtain confidence limits which can be considered as exact for Type II censoring. Dumonceaux (1969) prepared tables for confidence limits based on the MLEs. They were reproduced by Bain (1978). Schmee, Gladstein, and Nelson (1985) gave more extensive tables than those of Dumonceaux for the MLEs.

We notice that the tables were generated using the original (biased) MLEs. Hence, when confidence intervals are computed using these tables the biased MLEs have to be used. Strictly speaking these simulated confidence limits are valid only for Type II censored samples. But if the sample size is not too small they might also be used to obtain approximate confidence intervals for Type I censored samples, as suggested by Schmee et al. (1985), although more studies have to be done to see how the performance of this approximation depends on the sample size and degree of censoring. The tables are reproduced below as Tables 6.3 and 6.4. To demonstrate how the tables are used to obtain confidence limits we consider again Example 4.2.

Example 6.1

The data are from Gupta (1952) and give the survival times of 7 out of 10 mice. The MLEs computed in Example 4.7 are $\hat{\mu} = 1.742$ and $\hat{\sigma} = 0.792$. The $1 - \alpha = 95\%$ confidence limits for μ based on Table 6.3 are

$$\mu_L = \hat{\mu} - \hat{t}(1 - \alpha/2; n, r)\hat{\sigma} = 1.742 - 0.76 \quad 0.0792 = 1.68$$

$$\mu_U = \hat{\mu} - \hat{t}(\alpha/2; n, r)\hat{\sigma} = 1.742 - (-1.12) \quad 0.0792 = 1.83$$

Confidence limits for σ based on Table 6.4 are computed by

$$\sigma_L = \hat{\sigma}/\hat{w}(1 - \alpha/2; n, r) = 0.0792/1.45 = 0.054$$

$$\sigma_U = \hat{\sigma}/\hat{w}(\alpha/2; n, r) = 0.0792/0.392 = 0.20$$

Nelson and Schmee (1979) considered confidence limits for the BLUEs and prepared tables for sample sizes up to 10.

TABLE 6.3 Percentiles $\hat{t}(\delta; n, r)$ for Confidence Limits for μ

		δ						
r	n	.025	.05	.10	.50	.90	.95	.975
2	5	-41.2	-19.6	-9.29	-.79	.63	1.08	1.75
	6	-51.7	-23.9	-11.5	-1.04	.50	.79	1.22
	7	-55.1	-26.4	-12.6	-1.16	.42	.65	.90
	8	-60.6	-30.5	-14.1	-1.34	.37	.60	.79
	10	-70.6	-34.3	-15.9	-1.61	.31	.53	.68
	12	-75.6	-37.8	-18.4	-1.80	.27	.50	.66
	15	-85.8	-40.3	-19.6	-2.01	.26	.50	.65
	20	-100	-48.2	-22.4	-2.31	.22	.48	.65
	30	-112	-54.8	-25.9	-2.80	.20	.49	.70
	50	-123	-64.1	-30.9	-3.23	.21	.56	.78
3	5	-5.11	-3.38	-2.09	-.19	.73	1.10	1.55
	6	-5.61	-3.71	-2.28	-.24	.61	.89	1.25
	7	-6.51	-4.31	-2.63	-.32	.51	.75	1.01
	8	-7.10	-4.58	-2.82	-.39	.46	.65	.85
	10	-8.21	-5.55	-3.39	-.49	.40	.57	.71
	12	-9.82	-6.43	-3.91	-.60	.35	.51	.64
	15	-10.7	-7.05	-4.32	-.68	.33	.49	.60
	20	-12.7	-7.92	-4.84	-.81	.33	.50	.61
4	5	-2.16	-1.56	-1.06	-.05	.77	1.12	1.47
	6	-2.53	-1.81	-1.22	-.09	.63	.91	1.20
	7	-2.72	-1.95	-1.30	-.13	.56	.79	1.03
	8	-3.01	-2.21	-1.46	-.17	.48	.68	.88
	10	-3.59	-2.60	-1.73	-.24	.43	.58	.73
	12	-3.86	-2.78	-1.89	-.29	.38	.52	.64
	15	-4.61	-3.29	-2.18	-.36	.34	.48	.58
	20	-5.10	-3.68	-2.48	-.41	.34	.47	.57
5	6	-1.54	-1.17	-.83	-.03	.64	.90	1.16
	7	-1.67	-1.26	-.89	-.05	.57	.80	1.03
	8	-1.85	-1.35	-.95	-.08	.50	.69	.90
	10	-2.11	-1.56	-1.08	-.13	.43	.59	.75
	12	-2.33	-1.72	-1.21	-.16	.38	.51	.63
	15	-2.70	-2.01	-1.39	-.20	.35	.47	.58
	20	-3.19	-2.31	-1.60	-.27	.33	.44	.53
	30	-3.78	-2.80	-1.97	-.35	.33	.46	.55
	50	-4.62	-3.40	-2.41	-.47	.35	.50	.59
	70	-5.21	-3.84	-2.69	-.53	.37	.52	.64
	90	-5.47	-4.11	-2.85	-.55	.41	.57	.69
	100	-5.66	-4.21	-2.96	-.58	.42	.58	.70
6	7	-1.21	-.94	-.68	-.02	.59	.80	1.02
	8	-1.28	-.98	-.72	-.04	.52	.70	.89
	10	-1.46	-1.11	-.79	-.06	.44	.60	.75
	12	-1.58	-1.20	-.87	-.10	.39	.52	.65
	15	-1.86	-1.42	-1.02	-.13	.35	.46	.56
	20	-2.17	-1.65	-1.16	-.18	.32	.43	.51
7	8	-1.04	-.81	-.60	-.01	.53	.71	.88
	10	-1.12	-.88	-.64	-.04	.45	.61	.76
	12	-1.24	-.96	-.70	-.06	.40	.53	.66
	15	-1.40	-1.09	-.78	-.09	.34	.46	.56
	20	-1.63	-1.25	-.90	-.13	.31	.41	.49
8	10	-.94	-.74	-.55	-.02	.46	.61	.77
	12	-1.02	-.80	-.58	-.04	.40	.53	.66
	15	-1.13	-.88	-.65	-.06	.35	.46	.56
	20	-1.35	-1.03	-.74	-.09	.31	.41	.49
	30	-1.63	-1.25	-.91	-.14	.29	.38	.45
	50	-2.00	-1.56	-1.15	-.20	.30	.40	.47
10	12	-.78	-.62	-.47	-.01	.40	.53	.67
	15	-.83	-.66	-.49	-.03	.35	.47	.57
	20	-.95	-.75	-.55	-.05	.30	.39	.47
	30	-1.12	-.89	-.66	-.09	.27	.36	.42
	50	-1.41	-1.11	-.83	-.13	.28	.36	.43
	70	-1.65	-1.30	-.97	-.16	.29	.39	.47
	90	-1.80	-1.44	-1.07	-.19	.31	.41	.49
	100	-1.90	-1.45	-1.12	-.20	.33	.42	.50
12	15	-.68	-.55	-.41	-.02	.35	.46	.57
	20	-.75	-.60	-.44	-.03	.30	.39	.47
	30	-.86	-.68	-.51	-.06	.26	.33	.40
	50	-1.10	-.87	-.65	-.09	.25	.33	.39
15	25	-.62	-.49	-.37	-.03	.26	.34	.41
	35	-.69	-.55	-.41	-.04	.24	.31	.37
	50	-.81	-.66	-.49	-.06	.23	.30	.35
	70	-.94	-.75	-.57	-.08	.24	.32	.37
	90	-1.05	-.85	-.64	-.10	.26	.33	.40
	100	-1.12	-.89	-.67	-.10	.26	.34	.40
20	40	-.52	-.42	-.32	-.02	.22	.28	.34
	50	-.59	-.47	-.36	-.04	.21	.27	.32
	70	-.66	-.54	-.41	-.05	.21	.27	.32
	90	-.73	-.59	-.45	-.06	.22	.28	.34
	100	-.76	-.61	-.47	-.06	.23	.29	.34
25	35	-.41	-.34	-.27	-.01	.22	.29	.35
	50	-.45	-.37	-.28	-.02	.19	.25	.36
	70	-.51	-.42	-.32	-.03	.19	.24	.28
	90	-.57	-.46	-.35	-.04	.19	.25	.29
	100	-.59	-.48	-.37	-.04	.20	.25	.30
30	50	-.38	-.31	-.24	-.01	.19	.24	.28
	70	-.42	-.34	-.26	-.02	.17	.23	.27
	90	-.45	-.37	-.29	-.03	.18	.23	.27
	100	-.47	-.39	-.30	-.03	.17	.23	.27
40	50	-.32	-.26	-.20	.00	.19	.24	.28
	70	-.32	-.27	-.21	-.01	.16	.21	.25
	90	-.34	-.28	-.22	-.02	.16	.20	.24
	100	-.35	-.29	-.22	-.02	.16	.20	.24
50	70	-.27	-.23	-.18	-.00	.16	.20	.24
	90	-.28	-.23	-.18	-.01	.15	.19	.22
	100	-.29	-.24	-.19	-.01	.14	.18	.22

Note: r = number of failures; n = total number of units of test; δ = percentile.

Source: From Schmee, Gladstein, and Nelson (1985).

TABLE 6.4 Percentiles $\hat{w}(\delta; n, r)$ for Confidence Limits for σ

		δ						
r	n	.025	.05	.10	.50	.90	.95	.975
2	4.5	.017	.036	.071	.42	1.14	1.39	1.64
	10	.015	.032	.064	.39	1.13	1.40	1.66
	50	.015	.029	.058	.37	1.11	1.41	1.68
3	4	.12	.17	.24	.66	1.24	1.42	1.58
	5	.11	.16	.23	.63	1.24	1.42	1.61
	20	.10	.14	.21	.60	1.24	1.46	1.65
4	5	.22	.28	.36	.75	1.25	1.41	1.55
	10	.19	.24	.33	.72	1.26	1.43	1.59
	20	.18	.24	.32	.71	1.27	1.45	1.62
5	6	.29	.36	.44	.81	1.26	1.39	1.51
	10	.27	.33	.41	.78	1.26	1.41	1.55
	20	.25	.32	.40	.77	1.26	1.42	1.58
	50	.25	.31	.38	.76	1.28	1.45	1.60
	100	.24	.30	.38	.76	1.29	1.47	1.63
6	7	.36	.42	.50	.84	1.25	1.37	1.47
	10	.34	.40	.48	.83	1.25	1.39	1.51
	20	.32	.38	.46	.82	1.26	1.40	1.53
7	8	.41	.47	.55	.86	1.24	1.35	1.45
	10	.38	45	.53	.85	1.24	1.37	1.47
	20	.37	.43	.51	.85	1.26	1.38	1.50
8	9	.45	51	59	.89	1.23	1.33	1.42
	10	.43	.49	57	.88	1.23	1.34	1.44
	20	.41	.47	54	.87	1.25	1.36	1.46
	50	.39	.45	53	.86	1.26	1.39	1.50

		δ						
r	n	.025	.05	.10	.50	.90	.95	.975
10	12	.50	.56	.63	.91	1.22	1.31	1.40
	20	.48	.54	.61	.90	1.23	1.33	1.42
	50	.46	.52	.60	.89	1.24	1.35	1.45
	100	.45	.50	.58	.88	1.25	1.37	1.47
12	15	.55	.60	.67	.92	1.21	1.29	1.37
	20	.53	.59	.65	.92	1.22	1.30	1.38
	50	.51	.57	.64	.91	1.23	1.33	1.42
15	20	.59	.64	.71	.94	1.19	1.27	1.34
	50	.57	.62	.68	.93	1.21	1.29	1.37
	100	.55	.60	.67	.92	1.22	1.32	1.40
20	30	.65	.69	.75	.95	1.18	1.24	1.31
	50	.62	.67	.73	.95	1.18	1.25	1.32
	100	.61	.66	.72	.95	1.19	1.27	1.34
25	30	.69	.74	.78	.96	1.16	1.22	1.26
	50	.68	.72	.77	.96	1.17	1.23	1.29
	100	.66	.70	.76	.95	1.18	1.24	1.30
30	50	.71	.75	.80	.97	1.15	1.21	1.25
	100	.69	.73	.78	.96	1.16	1.22	1.27
50	70	.78	.81	.85	.98	1.12	1.16	1.20
	100	.77	.80	.84	.98	1.13	1.17	1.21
70	100	.82	.84	.87	.98	1.10	1.14	1.17
90	100	.85	.87	.89	.99	1.09	1.12	1.14

Note: r = number of failures; n = total number of units on test; δ = percentile.

Source: From Schmee et al., (1985).

Example 6.2

We now consider large sample confidence intervals for the data given in Example 6.1. In Example 4.7 we computed the MLEs of the log data which were $\mu = 1.742$ and $\hat{\sigma} = 0.792$. Applying the bias correction (Example 4.9) we obtained the corrected estimates 1.745 and 0.892. We calculate the confidence limits for the mean. The z-statistic (6.33) and (6.34) yields the 95% confidence interval (1.68, 1.83), which agrees with the simulated confidence interval given by Schmee, Gladstein, and Nelson (1985). Notice that the confidence interval is not symmetric about μ. We also mention that a direct comparison of the factors (6.41) and (6.42) with the simulated factors of Schmee et al. is not possible, since (6.41) and (6.42) are based on the bias-corrected MLEs while the simulated factors are based on the biased MLEs. But as the bias diminishes the factors agree. The 95% confidence interval for σ based on (6.47) and (6.48) is (0.0546, 0.1717), which is much closer to the confidence limit obtained by Table 6.4 than the log-transformed confidence interval (0.045, 0.14) given by (6.44) and (6.45).

Confidence Limits for Percentiles

The percentiles of the normal distribution are given as

$$y_p = \mu + u_p \sigma$$

where u_p is the 100pth percentile of the standard normal distribution. The percentiles of the lognormal distribution are then given by antilog(y_p). For instance the median of the lognormal distribution is antilog(μ).

For sample sizes up to 10 simulated confidence limits for the percentiles are given by Nelson and Schmee (1979) and reproduced by Nelson (1982). They can be considered exact for Type II censored samples and they are based on the BLUEs. For Type I censored samples they are approximate confidence intervals.

Nelson (1982) also gives approximate confidence intervals based on the F distribution and BLUEs. Nelson (1982) mentions that Mann, Schafer, and Singpurwalla (1974) have stated that this approximation works satisfactorily for $p \leq 0.1$ and large numbers of uncensored observations.

Large Sample Confidence Intervals

Since $\hat{\mu} + z\hat{\sigma}$ is asymptotically normal distributed we can obtain large sample confidence intervals using the methodology of Subsection B. We consider the statistic

$$T = \hat{\mu} + u_p\hat{\sigma} \pm z\hat{\sigma} \tag{6.49}$$

The asymptotic expectation and variance of this statistic are

$$E[T] = \mu + (u_p \pm z)\sigma \tag{6.50}$$

and

$$\text{var}(T_p) = \sigma^2 D/n \tag{6.51}$$

respectively, where D is

$$D = [\gamma_{11} + (u_p \pm z)^2\gamma_{22} + 2(u_p \pm z)\gamma_{12}] \tag{6.52}$$

and γ_{ij} are the variance factors tabulated in Appendix A.

We now determine z_1 and z_2 such that

$$P(\mu + u_p\sigma + z_1\sigma < \mu + u_p\sigma) = \alpha/2 \tag{6.53}$$

and

$$P(\hat{\mu} + u_p\sigma - z_2\hat{\sigma} > \mu + u_p\hat{\sigma}) = \alpha/2 \tag{6.54}$$

Using the asymptotic normality of the statistic T, we find that z_1 and z_2 are solutions to

$$\Phi[-z/\text{var}(T)] = \alpha/2 \tag{6.55}$$

and thus we obtain

$$z_1 = -A + \sqrt{(A^2 + B)} \tag{6.56}$$

and

$$z_2 = A + \sqrt{(A^2 + B)} \tag{6.57}$$

where

$$A = -u^2_{\alpha/2}(u_p\gamma_{22} + \gamma_{12})/[n - \gamma_{22}u^2_{\alpha/2}] \tag{6.58}$$

and

$$B = -u^2_{\alpha/2}(\gamma_{11} + \gamma_{22}u^2_p + 2u_p\gamma_{12})/[n - \gamma_{22}u^2_{\alpha/2}] \tag{6.59}$$

Thus, large sample confidence limits for y_p are given by

$$y_{pL} = \hat{\mu} + u_p\hat{\sigma} + z_1\hat{\sigma} \tag{6.60}$$

$$y_{pU} = \hat{\mu} + u_p\hat{\sigma} - z_2\hat{\sigma} \tag{6.61}$$

Simulation results show that these large sample confidence intervals are sufficiently accurate for sample sizes 10 and more (see Schneider and Weissfeld, 1986).

If limits are calculated from the logs of lognormal data, the antilogs of the limits are confidence limits for the lognormal percentiles.

Example 6.3

We consider the locomotive control example (see Example 4.3). The example is from Nelson (1982). He considered 96 locomotive controls, 59 of which were Type I censored. The MLEs of the log (basis 10) data, computed in Example 4.3 are $\hat{\mu} = 2.222$ and $\hat{\sigma} = 0.306$. Using the bias correction we obtain $\hat{\mu} = 2.228$ and $\hat{\sigma} = 0.314$. The 95% confidence limits for μ based on (6.33) and (6.34) are $\mu_L = 2.148$ and $\mu_u = 2.338$. These confidence limits agree with the confidence limits obtained by four-way interpolation in Table 6.3 (see Schmee et al., 1985).

The 95% confidence limits for σ based on (6.47) and (6.48) are $\sigma_L = 0.246$ and $\hat{\sigma}_u = 0.415$, which are closer to the confidence limits given by Table 6.4, namely, (0.24, 0.42), than the log-transformed confidence limits (0.24, 0.40).

The 10th percentile is $y_{0.1} = 2.228 + u_{0.1}0.314 = 1.826$. The confidence interval for the 10th percentile computed by (6.60) and

(6.61) is (1.718, 1.905). We finally compute the median and 10th percentile life with the confidence intervals by taking the antilogs. In thousand hours, the median with confidence interval is (141, 169, 218), and the 10th percentile life is (52, 67, 80).

Confidence Intervals for the Parameters of a Linear Model

Confidence limits for the parameters of a linear model can be derived in a way similar to that described in the sections above. In literature symmetrical confidence intervals are often computed using the asymptotic normality of the t-statistic. For small to moderate sample sizes, this method is reliable only for symmetrical censoring. For singly censored samples the confidence intervals should be unsymmetrical about β. Further research has to be done to obtain confidence intervals which are applicable for moderate sample sizes.

REFERENCES

Bain, L. J. (1978). Statistical Analysis of Reliability and Life-Testing Models, New York, Marcel Dekker.

Barr, D. R. and Davidson, T. (1973). A Kolmogorov-Smirnov test for censored samples, Technometrics 15, 739-757.

Chernoff, H., Gastwirth, J. L., and Johnes, M. V. (1967). Asymptotic distribution of linear combinations of functions of order statistics with applications to estimation, Annals of Mathematical Statistics 38, 52-72.

Dumonceaux, R. H. (1969). Statistical inference for location and scale parameters distributions, unpublished Ph.D. dissertation, University of Missouri-Rolla, Department of Mathematics and Statistics.

Filliben, J. J. (1975). The probability plot correlation coefficient test for normality, Technometrics 17, 111-117.

Gerlach, B. (1980). A correlation-type goodness-of-fit test for normality with censored sampling, Mathematische Operationsforschung und Statistik 11, 207-218.

Hall, I. J. and Sampson, C. B. (1973). One-sided tolerance limits for a normal population based on censored samples, Journal of Statistical Computation and Simulation 2, 317-324.

Johnson, N. L. and Kotz, S. (1970). Continuous Distributions, Vol. 1, New York, Houghton-Mifflin.

Kalbfleisch, J. D. and Prentice, R. L. (1980). The Statistical Analysis of Failure Time Data, New York, Wiley.

Lurie, D., Hartley, H. O., and Stroud, M. R. (1974). A goodness-of-fit test for censored data, Communications in Statistics 3, 745-753.

Mann, N. R., Schafer, R. E., and Singpurwalla, N. D. (1974). Methods for Statistical Analysis of Reliability and Life Data, New York, John Wiley.

Mehrotra, K. G. (1981). On goodness-of-fit tests based on spacing for Type II censored samples, Communications in Statistics 11, 869-878.

Mihalko, D. P. and Moore, D. S. (1980). Chi square tests of fit for Type II censored data, Annals of Statistics 8, 625-644.

Nelson, W. and Schmee, J. (1979). Inference for (log) normal life distributions from small singly censored samples and BLUEs, Technometrics 21, 43-54.

Nelson, W. (1982). Applied Life Data Analysis, New York, Wiley.

Neyman, J. and Pearson, E. S. (1928). On the use and interpretation of certain test criteria for purposes of statistical inference, Biometrika 20A, 175-240 and 263-294.

Patnaik, P. B. (1949). The non-central and F distribution and their applications, Biometrika 36, 202-232.

Pettitt, A. N. (1976). Cramer-von Mises statistics for testing normality with censored samples, Biometrika 63, 475-481.

Pettitt, A. N. and Stephens, M. A. (1976). Modified Cramer-von Mises statistics for censored data, Biometrika 63, 291-298.

Plackett, R. L. (1958). Linear estimation from censored data, Annals of Mathematical Statistics 29, 131-142.

Rao, C. R. (1947). Large sample tests of statistical hypotheses concerning several parameters with applications to problems of estimation, Proceedings of the Cambridge Philosophical Society 44, 50-57.

Rao, K. C. and Robson, D. S. (1974). A chi-square statistic for goodness-of-fit within the exponential family, Communications in Statistics 3, 1139-1153.

Ryan, T. and Joiner, B. (1973). Normal probability plots and tests for normality, Penn State Technical Report.

Samford, M. R. and Taylor, J. (1959). Censored observations in randomized block experiments, Journal of the Royal Statistical Society 21B, 214-237.

Saw, J. G. (1961). The bias of the maximum likelihood estimates of location and scale parameters given a Type II censored normal sample, Biometrika 48, 448-451.

Schmee, J., Gladstein, D., and Nelson, W. (1985). Confidence limits for parameters of a normal distribution from singly censored samples using maximum likelihood, Technometrics 27, 119-128.

Schneider, H. and Weissfeld, L. (1986). Inference based on Type II censored samples, Biometrics, to appear.

Serfling, R. J. (1980). Approximation Theorems of Mathematical Statistics, New York, Wiley.

Shapiro, S. S. and Brain, C. W. (1981). A review of distributional testing procedures and development of a censored sample distributional test, in Statistical Distributions in Scientific Work, C. Tailie, G. P. Patil, and B. A. Baldessari, eds., Vol. 5, 1-24.

Shapiro, S. S. and Wilk, M. B. (1965). An analysis of variance test for normality, Biometrika 52, 591-611.

Smith, R. M. and Bain, L. J. (1976). Correlation type goodness-of-fit statistics with censored sampling, Communication in Statistics, Theory and Methods A5, 119-132.

Tiku, M. L. (1967). Estimating the mean and standard deviation from a censored normal sample, Biometrika 54, 155-165.

Tiku, M. L. (1971). Estimating the mean and standard deviation from two censored normal samples, Biometrika 58, 241-243.

Tiku, M. L. (1973). Testing group effects from Type II censored normal samples in experimental design, Biometrics 29, 25-33.

Tiku, M. L. (1974a). A new statistic for testing normality, Communications in Statistics 3, 223-232.

Tiku, M. L. (1974b). Testing normality and exponentiality in multi-sample situations, Communications in Statistics 3, 777-794.

Tiku, M. L. (1975). A new statistic for testing suspected outliers, Communications in Statistics 4, 737-752.

Tiku, M. L. (1977). Rejoinder: Comment on "A new statistic for testing suspected outliers," Communications in Statistics A6, 1417-1422.

Tiku, M. L. (1978). Linear regression model with censored observations, Communications in Statistics, Theory and Methods A7(13), 1219-1232.

Tiku, M. L. (1981). Robust two-sample test and robust regression and analysis-of-variance via MML estimators, in Statistics and Related Topics, North-Holland Publishing Company, 297-314.

Tiku, M. L. (1982). Robust statistics for testing equality of mean or variances, Communications in Statistics, Theory and Methods 11, 2543-2558.

Tiku, M. L. and Singh, M. (1982). Robust statistics for testing mean vectors of multivariate distributions, Communications in Statistics, Theory and Methods 11, 985-1001.

Tiku, M. L. and Stewart, D. (1977). Estimating and testing group effects from Type I censored normal samples in experimental design, Communications in Statistics, Theory and Methods A6, 1485-1501.

Verill, S. P. (1981). Some asymptotic results concerning censored data versions of the Shapiro-Wilk goodness-of-fit test, Ph.D. dissertation, University of Wisconsin-Madison.

Verrill, S. P. and Johnson, R. A. (1983). The asymptotic distributions of censored data versions of the Shapiro-Wilk test of normality statistic, Technical Report No. 702, University of Wisconsin.

Wald, A. (1943). Tests of statistical hypotheses concerning several parameters when the number of observations is large, Transactions of the American Mathematical Society 54, 426-482.

Wilson, E. B. and Hilferty, M. M. (1931). The distribution of chi-square, Proceedings of the National Academy of Sciences, Washington, 17, 684-688.

CHAPTER 7

Variable Sampling Plans

In this chapter we shall consider some test problems. Truncated normal distributions appear frequently in quality control problems. If the quality characteristic under consideration is a quantitative variable such as weight or length of a product, variable sampling plans have a lower sample size than attributive sampling plans given certain risks (α, AQL), (β, LQL). Often, in order to meet consumers' specifications, the supplier uses some techniques to ensure that the quality characteristic lies between limits x_ℓ and x_r. These limits are not necessarity the specification limits assigned by the consumer. As a result of such screening methods a truncated normal distribution would be appropriate provided the quality characteristic before screening was normally distributed.

In Section 7.1 we shall consider variable sampling plans, given that the sample size is large enough to apply asymptotic theory.

7.1 VARIABLE SAMPLING PLANS FOR TRUNCATED NORMAL POPULATIONS

Most of the published sampling plans for inspection by variables, e.g., the MIL-STD-414 and BS 6002, assume a normal distribution of the inspected quality characteristics. It is often the case that this assumption is not justifiable, since there are some departures from the normal distribution. Some investigations concern the robustness of the variable sampling plans if the skewness and the excess differ from normality. A summary of work on variable acceptance sampling with emphasis on non-normality is given by Owen (1969). Das and Mitra (1964) use the Cornish-Fisher approximation up to three terms to compute the corresponding probabilities, i.e., those of rejecting lots with an acceptable quality level (AQL) and of accepting lots which have limiting quality (LQL). A non-normal distribution with known skewness and excess is assumed. Masuda (1978), using simulation, investigates the robustness of normal sampling plans applied to student and lognormal distributions. Schneider and Wilrich (1981) investigate the operating characteristic curve of a variable sampling plan when the underlying distribution is the beta-distribution. Furthermore, it was pointed out that in the case of a beta-distribution the Cornish-Fisher approximation is rather poor, since the probabilities of acceptance computed by this method are much too high.

Rao, Subrahmaniam, and Owen (1972) have studied the effect of non-normality on tolerance limits via the Edgeworth expansion. For the case in which the quality characteristic is non-normally but symmetrically distributed with thick tails, Kocherlakota and Balakrishnan (1984) suggested using symmetric Type II censoring to obtain robust estimates of μ and σ. In this section the effect of truncation (Schneider, 1985) is investigated.

A truncated normal distribution is suitable in many practical situations of acceptance sampling where there is a restriction on the variable under consideration. For instance, a truncated normal distribution arises in production engineering when sorting procedures eliminate items above or below designated tolerance limits. Truncation often results from technical constraints of a production process.

Variable Sampling Plans for Known σ^2

First we will investigate the robustness of standard variable sampling plans for known σ^2 with respect to the truncation points. Afterwards we will deal with the design of variable sampling plans based on a truncated normal distribution with given variance σ^2. The case of unknown variance is discussed at the end of the section.

<u>Robustness of variable sampling plans.</u> We assume the quality characteristic of the inspected item to be a normally distributed random variable X with parameters μ, σ^2, and truncation point x_r, where σ^2 and x_r are known. A one-sided specification limit is assigned. Let us consider an upper limit $T_u \leq x_r$. Items which have $X > T_u$ are nonconforming. Note that a lower limit T_l can be treated similarly. The pdf of the quality characteristic X is given as

$$f(x; \mu, \sigma, x_r) = \begin{cases} \phi\left(\frac{x-\mu}{\sigma}\right) \Big/ \left[\sigma\Phi\left(\frac{x_r-\mu}{\sigma}\right)\right] & x \leqslant x_r \\ 0 & x > x_r \end{cases} \tag{7.1}$$

where $\phi(\cdot)$ and $\Phi(\cdot)$ are the standard normal pdf and cdf, respectively. The proportion of nonconforming items in a lot is thus

$$p = \left[\Phi\left(\frac{x_r-\mu}{\sigma}\right) - \Phi\left(\frac{T_u-\mu}{o}\right)\right] \Big/ \Phi\left(\frac{x_r-\mu}{\sigma}\right)$$

Let $u_p^* = (T_u - \mu)/\sigma$ and $\Delta = (x_r - T_u)/\sigma$, then we obtain

$$p = 1 - \Phi(u_p^*)/\Phi(\Delta + u_p^*) \tag{7.2}$$

According to an agreement between the producer and the consumer, lots with a fraction nonconforming $p \leqslant$ AQL are presumed to be good and ought to be accepted with probability at least $1 - \alpha$. Furthermore, lots with $p >$ LQL are not acceptable to the customer and should be rejected with probability at least $1 - \beta$.

Lieberman and Resnikoff (1955) established the following procedure of variables acceptance sampling for the non-truncated case. If the specification is an upper limit, the value t

$$t = \overline{x} + k\sigma \tag{7.3}$$

of the test statistic $T = \overline{X} + k\sigma$ is compared with the specification limit T_u. On the basis of this comparison, each lot is either accepted ($t \leqslant T_u$) or rejected ($t > T_u$). Accordingly, a variable plan is specified by the parameters n (sample size) and k (acceptance constant). The desired (n, k)-plan has to fulfill the condition that the OC-curve of the plan will pass through the points defined by (AQL, $1 - \alpha$) and (LQL, β). To compute the sample size n and the acceptance constant k, we have to analyze the distribution of T for a given fraction nonconforming p. The OC-curve is given as

$$L(p) = P(T \leqslant T_u | p) \tag{7.4}$$

If the variance of X is known the statistic T is normally distributed. Taking into consideration the requirements

$$L(AQL) = P(T \leq T_u | AQL) = 1 - \alpha \tag{7.5}$$

$$L(LQL) = P(T \leq T_u | LQL) = \beta \tag{7.6}$$

it is not difficult to calculate the parameters (n, k). There are well-known published sampling plans such as MIL-STD-414 and British STD (BS 6002). In order to investigate the robustness of these sampling plans with respect to a truncation, the OC-curve of these plans is examined for the case where X is truncated at x_r. Given a truncation point x_r the expected value and the variance of T are

$$E[T] = \mu - \sigma W(u_r) + k\sigma \tag{7.7}$$

and

$$\mathrm{var}[T] = \frac{\sigma^2}{n}\{1 - W(u_r)[W(u_r) + u_r]\} \tag{7.8}$$

respectively, where (see Chapter 2)

$$W(u_r) = \phi(u_r)/\Phi(u_r) \tag{7.9}$$

$$u_r = (x_r - \mu)/\sigma = \Delta + u_p^* \tag{7.10}$$

The OC-curve of an (n, k)-plan is then asymptotically given by

$$L(p) = \Phi\left\{[u_p^* - k + W(\Delta + u_p^*)]\sqrt{n\gamma(\Delta + u_p^*)}\right\} \tag{7.11}$$

where

$$\gamma(\Delta + u_p^*) = \{1 - W(\Delta + u_p^*)[W(\Delta + u_p^*) + \Delta + u_p^*]\}^{-1} \tag{7.12}$$

and u_p^* is the solution to Eq. (7.2).

Example 7

We consider the example AQL = 0.01, LQL = 0.03, $\alpha = 0.1$, and $\beta = 0.1$. In the untruncated case we obtain the sampling plan n = 34, k = 2.106. Applying this plan to a truncated normal distribution the probabilities of acceptance alter according to the amount of truncation.

For small values Δ the probabilities of acceptance differ significantly from 1 - α and 1- β, respectively. This means for instance that if a producer screens his production near the upper specification limit and the consumer applies a standard variable sampling plan it is very likely that the lot will be rejected.

Figure 7.1 shows the operating characteristic curve (OC) for Example 7.1. The upper curve gives the OC of an (n, k) sampling plan for a normally distributed quality characteristic, while the dashed line shows the OC of the same plan for a truncated normal population.

The design of sampling plans for the singly truncated case. We consider the statistic

$$T' = \mu_{ML} + k\sigma$$

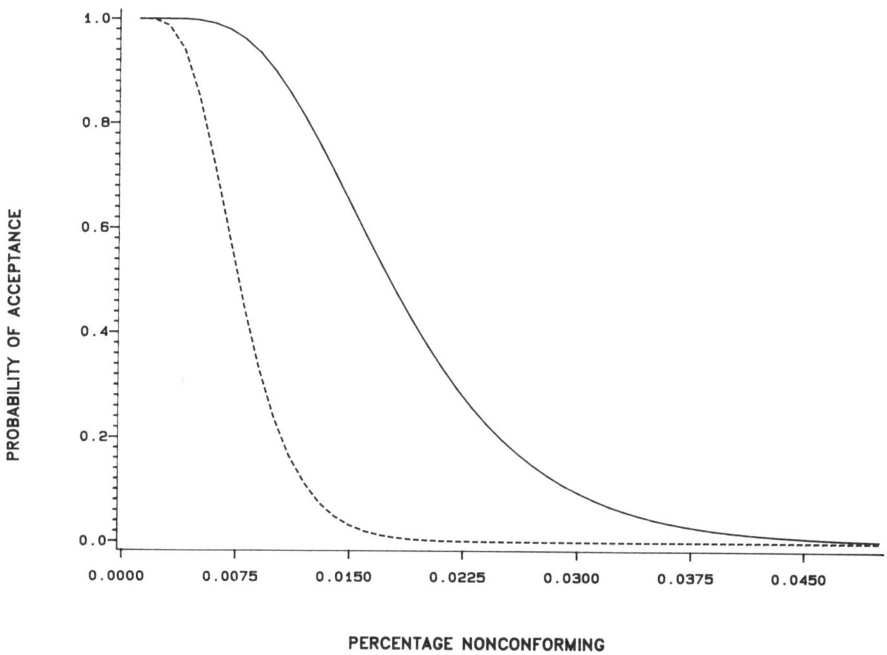

FIGURE 7.1 OC-curve of plan (34, 2.106). ____, untruncated. ---- , truncated (Δ = 0.2).

where μ_{ML} is the maximum likelihood estimator discussed in Section 3.1. Since σ^2 is assumed to be known the ML-estimate is

$$\mu_{ML} = \bar{x} + \delta\sigma$$

where δ is the solution to $\delta = W(\theta - \delta)$ and $\theta = (x_r - \bar{x})/\sigma$. We propose that one of the solution methods given in Section 3.1 be used. The value

$$t' = \mu_{ML} + k\sigma$$

is compared with the upper specification limit T_u. On the basis of this comparison each lot is either accepted or rejected. In order to determine a sampling plan such that its OC curve passes through the points (AQL, $1 - \alpha$) and (LQL, β), the distribution of T' has to be evaluated. From Chapter 3 we know that μ_{ML} is asymptotically normal with mean μ and variance $(\sigma^2/n)\gamma(u_r)$ where $\gamma(u_r)$ is given by

$$\gamma(u_r) = \{1 - W(u_r)[W(u_r) + u_r]\}^{-1} \tag{7.13}$$

Hence the OC-curve is approximated by

$$L(p) = P(T' \leq T_u | p) = \Phi\{(u_p^* - k)[n/\gamma(u_r)]^{1/2}\} \tag{7.14}$$

where $u_r = \Delta + u_p^*$. n and k are solutions to

$$L(AQL) = 1 - \alpha \tag{7.15}$$

$$L(LQL) = \beta \tag{7.16}$$

Using (7.14) we are able to write (7.15) and (7.16) in the form

$$(u_{AQL}^* - k)[n/\gamma(u_{AQL}^* + \Delta)]^{1/2} = u_{1-\alpha} \tag{7.17}$$

and

$$(u_{LQL}^* - k)[n/\gamma(u_{LQL}^* + \Delta)]^{1/2} = u_\beta \tag{7.18}$$

where u^*_{AQL} and u^*_{LQL} are solutions to (7.2) and $u_{1-\alpha}$ and u_β are the quantiles of a standard normal distribution. And thus

$$n^* = \frac{[u_{1-\alpha}\gamma(u^*_{AQL} + \Delta)^{1/2} - u_\beta\gamma(u^*_{LQL} + \Delta)^{1/2}]^2}{(u^*_{AQL} - u^*_{LQL})^2} \tag{7.19}$$

$$k^* = u^*_{AQL} - u_{1-\alpha}[\gamma(u^*_{AQL} + \Delta)/n^*]^{1/2} \tag{7.20}$$

where n^* has to be rounded off.

Since the derivation of these sampling plans is based on asymptotic results a Monte Carlo simulation was performed by Schneider (1985). This study shows that in all cases the OC-curve goes through (AQL, 0.9) as required, but the probability of acceptance at LQL is in some cases lower than 0.1. The latter is the case for small sample sizes and small Δ.

Note that there is a maximum percentage of defectives in a lot for given Δ. The percentage of defectives takes its maximum at $\mu = (x_r + T_u)/2$; hence

$$p_{max} = 2 - 1/\left[\Phi\left(\frac{\Delta}{2}\right)\right]$$

The Doubly Truncated Normal Distribution

If the normal distribution is truncated at both sides, variable sampling plans can be designed in a similar way to the single truncated case. The pdf is

$$f(x;\mu, \sigma, x_r, x_\ell) = \begin{cases} \sigma^{-1}\varphi\left(\frac{x-\mu}{\sigma}\right) \Big/ \left[\Phi\left(\frac{x_r-\mu}{\sigma}\right) - \Phi\left(\frac{x_\ell-\mu}{\sigma}\right)\right] \\ x_\ell \leqslant x \leqslant x_r \\ 0 \qquad \text{otherwise} \end{cases} \tag{7.21}$$

Given an upper specification limit T_u, the percentage of defective items in a lot is

$$p = [\Phi(\Delta_\ell + u^*_p) - \Phi(u^*_p)]/[\Phi(\Delta_r + u^*_r) - \Phi(\Delta_\ell + u^*_p)] \tag{7.22}$$

where

$$\Delta_r = (x_r - T_u)/\sigma, \quad \Delta_\ell = (x_\ell - T_u)/\sigma, \quad u_p^* = (T_u - \mu)/\sigma \tag{7.23}$$

Using the statistic $T' = \hat{\mu} + k\sigma$, we have to solve the equation

$$\delta = Z_r - Z_\ell \tag{7.24}$$

to obtain the ML-estimator μ_{ML} where

$$Z_r = \phi(u_r)/[\Phi(u_r) - \Phi(u_\ell)]$$

$$Z_\ell = \phi(u_\ell)/[\Phi(u_r) - \Phi(u_\ell)]$$

$u_\ell = (x_\ell - \mu)/\sigma$ and u_r is given by (7.10). The asymptotic mean and variance are given by

$$E[\mu_{ML}] = \mu \tag{7.25}$$

$$\mathrm{var}[\mu_{ML}] = \frac{\sigma^2}{n}\gamma' \tag{7.26}$$

where

$$\gamma' = [1 + u_\ell Z_\ell - u_r Z_r - (Z_\ell - Z_r)^2]^{-1} \tag{7.27}$$

Replacing γ by γ' in formulas (7.19) and (7.20) yields a variable sampling plan (n^{**}, k^{**}) in the double truncated case.

Variable Sampling Plans for Unknown σ^2

We consider a quality characteristic X which has a right truncated pdf as given by (4.1) in which σ^2 is assumed to be unknown. An upper specification limit $T_u \leqslant x_r$ is assigned. Let $u_p^* = (T_u - \mu)/\sigma$ and $\Delta = (x_r - T_u)/\sigma$; then we obtain as in (7.2)

$$p = 1 - \frac{\Phi(u_p^*)}{\Phi(\Delta + u_p^*)} \tag{7.28}$$

But now, since σ is unknown, there is no one-to-one relationship between u_p^* and p. We will proceed as in the known σ case and define the statistic

$$T = \mu_{ML} + k\sigma_{ML}$$

We know that T is asymptotically normally distributed with expected value

$$E[T] = \mu + k\sigma$$

and variance

$$\text{var}[T] = \text{var}[\mu_{ML}] + k^2 \text{var}[\sigma_{ML}] + 2k\,\text{cov}[\mu_{ML}, \,_{ML}]$$

where $\text{var}[\mu_{ML}]$, $\text{var}[\,_{ML}]$, and $\text{cov}[\mu_{ML}, \sigma_{ML}]$ were derived in Section 3.1.

Hence it is not difficult to find the distribution of T for large n, but the operating characteristic curve will unfortunately depend on the unknown σ. The OC-curve is defined by

$$L(p) = P(T \leqslant T_u | p)$$

Thus the OC-curve of an (n, k)-plan is then approximately

$$L(p) = \Phi\{[n/\gamma(\Delta + u_p^*, k)]^{1/2}[u_p^* - k]\} \tag{7.29}$$

where

$$\gamma(\Delta + u_p^*, k) = \frac{J_{22} + k^2 J_{11} - 2kJ_{12}}{J_{11}J_{22} - J_{12}J_{12}} \tag{7.30}$$

Since $\gamma(\Delta + u_p^*, k)$ is not only a function of p but also of Δ and thus of σ the OC-curve is dependent on the unknown σ^2.

A rough and ready method is to use s^2, the sample variance derived from the lot history, in order to determine a sampling plan for given risks (AQL, $1 - \alpha$) and (LQL, β). Such an approach would be justified if L(p) were not very sensitive to small changes of σ^2, given that the sample sizes are sufficiently large.

A simulation study (Schneider, 1985) shows that when the standard deviation differs about 10% from the assumed standard deviation, the probability of accepting a lot at the AQL level might decrease or increase about 10% if truncation is severe. If truncation is light to moderate the difference can be ignored for practical purposes.

7.2 SYMMETRICAL CENSORING TO OBTAIN ROBUST VARIABLE SAMPLING PLANS

In Section 7.1 we discussed the sensitivity of variable sampling plans with respect to truncation of the normal population. The results coincide with the findings of Schneider and Wilrich (1981), who have investigated the sensitivity of variable sampling plans for short tailed distributions. In this section we deal with outliers.

Although the elegance of the variables method and its efficiency when the assumption of normality is true and all observations are available makes this procedure superior to the attributes procedure, the sensitivity of the procedure to the assumption of normality leads to the attributes procedure being used even when the variables method can be applied. Farlie (1980) has described some undesirable features of acceptance sampling by variables that have hindered the widespread use of the method. Farlie (1980) gives an example to demonstrate the role outliers play in variable sampling plans. Consider an (n, k) sampling plan with $n = 3$ and $k = -1.12$. A lower limit $T_l = 0$ is specified. Hence, if $\bar{x} - 1.12s \geq 0$, the batch is accepted; otherwise, it is rejected. From two batches samples are taken where $x_1 = 0.15$, $x_2 = 1.15$, $x_3 = 2.15$ and $y_1 = 0.15$, $y_2 = 1.15$, $y_3 = 3.05$. The first sample leads to the acceptance of the associated batch because $\bar{x} = 1.15$, $s_x = 1$. The second sample leads to rejection of the associated batch since $\bar{y} = 1.45$ and $s_y = 1.473$. The result seems paradoxical since the y sample is intuitively better than the x sample, yet the better sample leads to the rejection of the batch and the poorer sample leads to acceptance of the batch. This paradox is caused by the large observation 3.05. The normality assumption translates the large value into evidence for large deviations in the other direction as well.

Censoring was proposed by Tiku (1980) as a robust test procedure. He showed that symmetrical Type II censoring seems to be a powerful method of obtaining robust estimation of the location parameter and performs quite well with other well-known robust estimators. As a conclusion of his findings Tiku (1980) states that

> Nonnormality essentially comes from the tails and once the extreme observations (representing the tails) are censored, there is hardly any difference between a normal sample and a nonnormal sample and in that situation whatever is good for normal is automatically good for nonnormal samples.

Kocherlakota and Balakrishnan (1984) used symmetrical censoring in order to obtain robust two-sided variable sampling plans. Although it seems unclear that symmetrical censoring will help to control the percentage nonconforming for arbitrary long tailed distributions, there are situations where this procedure can be successful. The problem is that we have to assume that the percentages of nonconforming above the upper specification limit and below the lower specification limit are given by the normal probabilities. If that is not the case, we cannot apply variables sampling plans because we do not know the relationship between the standardized mean and the percentage nonconforming. There are situations, however, where outliers do not change the tail probabilities. For the following we assume that the quality characteristic is normally distributed with parameters μ and σ, and an upper specification limit is specified. Outliers may occur in the left or right tail, but it is assumed that the percentage nonconforming is given by $p = 1 - \Phi[(T_u - \mu)/\sigma]$. Items which have $x > T_u$ are nonconforming. According to an agreement between the producer and consumer, lots with a fraction defectives $p \leq AQL$ are presumed to be good and ought to be accepted with probability at least $1 - \alpha$. Furthermore, lots with $p > LQL$ are not acceptable to the consumer and should be rejected with probability at least $1 - \beta$. Consider the procedure suggested by Lieberman and Resnikoff (1955). The value $t = x + ks$ is compared with the specification limit T_u. On the basis of this comparison, each lot is either accepted ($t \leq T_u$) or rejected ($t > T_u$). Accordingly, a variable sampling plan is specified by the parameters n (sample size) and k (acceptance constant). The desired (n, k)-plan has to fulfill the condition that the operating characteristic curve of the plan passes through the points defined by (AQL, $1 - \alpha$) and (LQL, β). We now derive the sample size and acceptance number for the case where the sample is symmetrically censored.

Given a sample of size n, the q percent smallest and q percent largest observations are censored. Let $n_\ell = n_r = qn$. We use the statistic

$$T = \hat{\mu} + k\hat{\sigma} \tag{7.31}$$

where $\hat{\mu}$ and $\hat{\sigma}$ are estimators of μ and σ in the censored case. The actual value $t = \hat{\mu} + k\hat{\sigma}$ is used as in the procedure of Lieberman and Resnikoff (1955) described above. The operating characteristic curve of this test is given by

$$L(p) = P(T \leqslant T_u | p) \tag{7.32}$$

Since the sample is symmetrically censored, estimators of the mean and standard deviation are uncorrelated. Furthermore the estimator of the mean will be unbiased while the most common estimators of the standard deviation are biased and a bias correction has to be used. The operating characteristic curve can then be written

$$L(p) = P\left(\frac{-\sqrt{n_0/\gamma}\,(\hat{\mu} - \mu)/\sigma + \sqrt{n_0/\gamma}\,(T_u - \mu)/\sigma}{\hat{\sigma}/\sigma} \geqslant \sqrt{n_0/\gamma}\,k\right) \tag{7.33}$$

where $n_0 = (1 - 2q)n$ and $\sigma^2\gamma/n_0$ is the asymptotic variance of $\hat{\mu}$. According to our assumption $(T_u - \mu)/\sigma = u_{1-p}$ where $u_{1-p} = \Phi^{-1}(1 - p)$. We notice that $u = -\sqrt{n_0/\gamma}\,[(\hat{\mu} - \mu)/\sigma]$ is asymptotically standard normally distributed and $\hat{\sigma}^2/\sigma^2$ is approximately χ^2 distributed with $f = n_0 - 1$ degrees of freedom. Hence

$$\frac{u + \delta}{\hat{\sigma}/\sigma} \tag{7.34}$$

follows approximately the noncentral t-distribution with $f = n_0 - 1$ degrees of freedom and noncentrality parameter $\delta = \sqrt{n_0/\gamma}\,u_{1-p}$.

As estimators for μ and σ, the maximum likelihood estimator, the modified maximum likelihood estimator of Tiku (1967), or the restricted maximum likelihood estimator can be used. But we notice that $\hat{\sigma}$ has to be bias corrected, i.e., we use, for instance, $\hat{\sigma} = \sigma_{ML}\sqrt{n_0/(n_0 - 1)}$. For $p_\ell = p_r = 0.1$ we obtain $\gamma = 0.828$.

In what follows we derive an approximation for the OC-curve which is useful to design (n, k)-plans without using tables of the noncentral t-distribution.

Consider the statistics

$$T = \hat{\mu} + k\hat{\sigma} \tag{7.35}$$

where $\hat{\mu}$ and $\hat{\sigma}$ have the required properties discussed above. Then

$$E[T] \approx \mu + k\sigma \tag{7.36}$$

and

$$\mathrm{var}[T] \approx \frac{\sigma^2}{n_0}\gamma + k^2\sigma^2\frac{1}{2f} \tag{7.37}$$

where $f = n_0 - 1$. Hence for large n

$$u = \frac{\hat{\mu} + k\hat{\sigma} - (\mu + k\sigma)}{\sigma[\gamma/n_0 + k^2/(2f)]^{1/2}} \tag{7.38}$$

is approximately standard normally distributed. The OC-curve is therefore approximately given by

$$L(p) = P[u \leqslant A(u_{1-p} - k)] = \Phi[A(u_{1-p} - k)] \tag{7.39}$$

where

$$A = \frac{1}{\sigma[\gamma/n_0 + k^2/(2f)]^{1/2}} \tag{7.40}$$

and

$$u_{1-p} = \frac{T_u - \mu}{\sigma} \tag{7.41}$$

We also can write (7.39) as

$$u_L - A(u_{1-p} - k) = 0 \tag{7.42}$$

where u_L corresponds to the probability of acceptance L.

Example

We want to determine a sampling plan the OC-curve of which goes through (0.01, 0.95) and (0.05, 0.10). In order to make the plan robust against outliers, 10% of the smallest and 10% of the largest observations are censored. Thus $\gamma = 0.828$ and we obtain the two equations

$$1.64 - \frac{1}{\sqrt{(0.828)/n_0 + k^2/[2(n_0 - 1)]}}(2.33 - k) = 0$$

$$-1.28 - \frac{1}{\sqrt{(0.828)/n_0 + k^2/(2(n_0 - 1)]}}(1.88 - k) = 0$$

We obtain $n_0 = 51$ and $k = 1.944$. The following sampling plan has therefore approximately the required risks: take a sample of size $n = 63$, censor the smallest 6 and the largest 6 observations, and calculate $\hat{\mu}$ and $\hat{\sigma}$. If $\mu + k\hat{\sigma} < T_u$ accept the lot; otherwise reject it.

In order to investigate the finite sample properties of these robust sampling plans, a simulation study is performed for some selected sampling plans. We have chosen L(AQL) = 0.95 and L(LQL) = 0.1 for various AQL and LQL values. Since the RML-estimator gives almost the same results as the ML-estimator we have used the former one because of its simplicity.

The simulation results presented in Table 7.1 show that also for finite sample sizes the true OC-curve meets the required risks quite well. In the last column the sample sizes of a standard variables sampling plan with the same risks are given. These sample sizes are obtained from a nomogram prepared by Wilrich (1970). We notice that the sample sizes n are nearly in agreement with n_0, the number of uncensored observations for the proposed robust variable sampling plan. Furthermore, the critical values k of he standard variable sampling plans do not differ from the k-value of the censored sampling plans in the first decimal place. Hence, a rough and ready method to obtain a robust sampling plan against outliers might be obtained by simply increasing the sample size of a standard variable sampling plan. We select a variable sampling plan (n, k) for given risks (ALQ, $1 - \alpha$) and (LQL, β) and increase the sample size to $n = n/(1 - 2q)$ where q gives the percentage of censoring at each side of the sample.

TABLE 7.1 Simulation Results for Robust Sampling Plan

n	n_0	k	AQL (%)	LQL (%)	$\hat{L}$(AQL)	$\hat{L}$(LQL)	n
73	59	2.186	0.5	3	0.95	0.11	60
49	39	2.259	0.3	3	0.94	0.12	40
39	31	2.317	0.2	3	0.94	0.11	32
64	52	1.947	1.0	5	0.94	0.11	52
38	30	2.055	0.5	5	0.94	0.11	30
30	24	2.145	0.3	5	0.94	0.10	24
23	19	2.187	0.2	5	0.94	0.12	20

Note: $\hat{L}$(AQL) and $\hat{L}$(LQL) are the means of 10,000 samples.

The RML-estimators μ_{RML} and $_{RML}$ are easy to compute and they are very close to the ML-estimates. We therefore give the explicit formulas for p = 10% and p = 5%. Assume the sample size of n is ordered and the n_ℓ = [qn] largest and n_r = [qn] smallest observations are censored. Let $n_0 = n - n_\ell - n_r$.

$$\bar{x} = \frac{1}{n_0} \sum_{i=n_\ell+1}^{n-n_r} x_{(i)} \tag{7.43}$$

$$s^2 = \frac{1}{n_0} \sum_{i=n_\ell+1}^{n-n_r} (x_{(i)} - \bar{x})^2 \tag{7.44}$$

then

$$C = e[\bar{x} - x_{(n_\ell+1)}] + h[x_{(n-n_r)} - x_{(n_\ell+1)}] \tag{7.45}$$

$$g = \frac{1}{2} C + \frac{1}{2} \left\{ C^2 + 4s^2 + 4[x - x_{(n_\ell+1)}]^2 \right\}^{1/2} \tag{7.46}$$

$$\mu_{RML} = \bar{x}$$

$$\sigma^2_{RML} = \frac{n_0}{n_0 - 1}(s^2 + g^2 f) \tag{7.48}$$

where e, h, and f are constants dependent only on q. For instance, if g = 0.05, then e = 1.645, h = 0.115, and f = 0.377; and if g = 0.1, then e = 1.282, h = 0.219, and f = 0.562.

7.3 VARIABLE SAMPLING PLANS FOR LOGNORMAL POPULATIONS

We shall now consider variable sampling plans with singly censored samples. The quality characteristic of the inspected item is assumed to be a lognormal distributed random variable y with parameter μ and σ. Consider for instance the lifetime of a product. A one-sided lower specification limit T_l is assigned. Items which have a lifetime $y < T_l$ are considered nonconforming. According to an agreement between the producer and consumer, lots with fraction nonconforming $p \leqslant$ AQL are presumed to be good and ought to be accepted with probability at least $1 - \alpha$. Furthermore, lots with $p >$ LQL are not acceptable to the consumer and should be rejected with probability at least $1 - \beta$. In order to save time and money product lifetime data can be analyzed before all test units have failed. We assume a sample of size n is chosen and started on test together until a prechosen number n_0 has failed. Thus, the $n - n_0$ observations are Type II censored from above. Some authors use the term failure censored (Nelson and Schmee, 1979).

Instead of using the actual lifetime y we use X = log(Y) which is normally distributed with mean μ and standard deviation σ. We notice that the percentiles of the lognormal distributions are given by $\exp(\mu + z_p\sigma)$ where z_p is the percentile of the standard normal distribution. We use the same procedure suggested by Lieberman and Resnikoff (1955) for normal data. Let $\hat{\mu}$ be an estimate of the mean and $\hat{\sigma}$ an estimate of the standard deviation of X, then the value

$$t = \hat{\mu} - k\hat{\sigma}$$

is compared with the logarithm of the lower limit $T_l' = \log(T_l)$.

On the basis of this comparison, each lot is either accepted ($t \geqslant T_l'$) or rejected ($t < T_l'$). A variable sampling plan which is specified by the sample size n and the acceptance constant k has to fulfill the condition that the operating characteristic curve passes through the points defined by (AQL, $1 - \alpha$) and (LQL, β). If we use asymptotically efficienc estimators of μ and σ, then the statistic $T = \hat{\mu} - k\hat{\sigma}$ has expectation

$$E[T] = \mu - k\sigma \tag{7.49}$$

and variance

$$\text{var}[T] = \frac{\sigma^2}{n_0}\left(\gamma_{11} + k^2\gamma_{22} - 2k\gamma_{12}\right) \tag{7.50}$$

where γ_{11}, γ_{22} and γ_{12} are the variance-covariance factors of the MLEs, which are given in Table A3 of Appendix A.

For large sample sizes n

$$u = \frac{\hat{\mu} - k\hat{\sigma} - (\mu - k\sigma)}{\sigma(\gamma_{11} + k^2\gamma_{22} - 2k\gamma_{12})^{1/2}}\sqrt{n_0} \tag{7.51}$$

is approximately normally distributed and thus the OC-curve is given by

$$L(p) = P[u \geq A(u_p + k)] = 1 - \Phi[A(u_p + k)]$$

where

$$A = \frac{\sqrt{n_0}}{(\gamma_{11} + k^2\gamma_{22} - 2k\gamma_{12})^{1/2}} \tag{7.53}$$

and

$$u_p = \frac{T_1 - \mu}{\sigma} \tag{7.54}$$

In order to determine an (n, k)-plan for given (AQL, $1 - \alpha$) and (LQL, β), we solve the equations

$$u_\alpha - A(u_p + k) = 0 \qquad (p = \text{AQL}) \tag{7.55}$$

$$u_{1-\beta} - A(u_p + k) = 0 \qquad (p = \text{LQL}) \tag{7.56}$$

for n_0 and k. Since n_0 has to be an integer value we choose $n_0^* = [n_0 + 1]$, where [x] is the integer part of x. The critical value k is then adjusted by solving equation (7.55) for $n_0^* = [n_0 + 1]$.

In order to facilitate the application of the sampling plans the RML or MML estimators can be used for estimating the mean and standard deviation of X. These estimators (see Chapter 4) are particularly easy to compute. It is also suggested to apply the bias corrections (4.178) and (4.179) in order to obtain nearly unbiased estimators of μ and σ.

In Table 7.2 some selected sampling plans are presented. The sampling plans are matched to the attributive sampling plans of MIL-STD-105D Code Letter K, so that they have approximately the same operating characteristic curve. We have chosen q = 30%, 50%, 70% censoring.

Example 7.2

We consider a batch of items the quality characteristics of which is the lifetime under stress. It is assumed to be known that the lifetime of this item is lognormal distributed. An item is conforming when its (accelerated) life is above $T_l = 1$ time units. In order to save time and money it was decided to choose a sampling plan with AQL = 0.1. The OC curve of this plan passes through (0.95, 0.00041) and (0.10, 0.0184). If a variable sampling plan without censoring is used, then n = 24 and k = 2.639 (see Wallis, 1947). If a plan with 70% censoring is used then a sample of size 50 is taken from the batch, and they are started on test together until the first 15 units have failed. The observed failure times of the first 15 units are:

1.6 2.0 3.3 3.3 3.4 3.5 3.8 4.4 4.5 4.5 4.8 5.0 5.8 5.9 5.9

Taking the logarithm on the basis 10 we obtain the sample mean $\bar{x}$ = 0.587 and the sample standard deviation $s^2 = 0.024$. Furthermore the logarithm of the largest observations is $\log_{10}[x_{(15)}] = 0.771$. The RML estimates, for instance, are $\hat{\mu} = 0.925$ and $\hat{\sigma} = 0.294$. Using the bias correction (4.178) and (4.179) we obtain the final estimates $\hat{\mu} = 0.942$ and $\hat{\sigma} = 0.3123$. We compute $t = \hat{\mu} + k\hat{\sigma} = 0.1156$. Since this value is larger than the specification limit log (1) we accept the batch. From the estimates we also calculate the median test time which is 5.9 time units. Using an uncensored sampling plan has the advantage of a lower sample size (n = 24) but the median test time (34.4) will be about six times as long as with the censored plan. A censored sampling plan therefore offers a considerable saving in test time. This has to be weighed against the increase in sampling.

TABLE 7.2 Selected Sampling Plans for Lognormal Populations, Matched with MIL-STD-105D Code Letter K ($\alpha = 0.05$, $\beta = 0.1$)

q	n	n_0	n_r	k	AQL (%)	p_α	p_β
30%	29	20	9	2.638	0.10	0.00041	0.01840
	43	30	13	2.257	0.40	0.00284	0.03110
	53	37	16	2.056	0.65	0.00654	0.04260
	59	41	18	1.911	1.00	0.01090	0.05350
	66	46	20	1.702	1.50	0.02090	0.07420
	70	49	21	1.550	2.50	0.03190	0.09420
50%	38	19	19	2.641	0.10	0.00041	0.01840
	54	27	27	2.259	0.40	0.00284	0.03110
	64	32	32	2.056	0.65	0.00654	0.04260
	68	34	34	1.909	1.00	0.01090	0.05350
	76	38	38	1.705	1.50	0.02090	0.07420
	78	39	39	1.552	2.50	0.03190	0.09420
70%	50	15	35	2.646	0.10	0.00041	0.01840
	67	20	47	2.262	0.40	0.00284	0.03110
	73	22	51	2.053	0.65	0.00654	0.04260
	77	23	54	1.908	1.00	0.01090	0.05350
	80	24	56	1.701	1.50	0.02090	0.07420
	80	24	56	1.549	2.50	0.03190	0.09420

Note: n = sample size; n_0 = number of uncensored observations; n_r = number of censored observations; $q = n_r/n$.

The sampling plans given in Table 7.2 are based on the asymptotic normality of the test statistic T. Monte Carlo simulations show, however, that the assigned risks are met even for small sample sizes.

REFERENCES

BS 6002 (1979). Specification sampling procedures and charts for inspection by variables for percent defectives, British Standards Institution, London.

Balakrishnan, N. and Kocherlakota, S. (1985). Robust two-sided tolerance limits based on MML estimators, Communication in Statistics, Theory and Methods 14, 175-184.

Das, N. G. and Mitra, S. K. (1964). The effect of non-normality on sampling inspection, Sankhya 261, 169-176.

Farlie, D. J. G. (1983). Sampling plans with property Q, in Frontiers in Statistical Quality Control, H. J. Lenz et al., eds., Physica-Verlag, Wurzburg, West Germany.

Hosono, Y., Okta, H., and Kase, S. (1981). Design of single sampling plans for doubly exponential characteristics, in Frontiers in Statistical Quality Control, H. J. Lenz et al., eds., Physica-Verlag, Wurzburg, West Germany, 94-112.

Kocherlakota, S. and Balakrishnan, N. (1984). Two-sided acceptance sampling plans based on MML estimators, Communication in Statistics, Theory and Methods 13, 3123-3131.

Lieberman, G. J. and Resnikoff, G. J. (1955). Sampling plans for inspection by variables, Journal of the American Statistical Association 50, 457-516.

Masuda, K. (1978). Effect of non-normality on sampling plans by Lieberman and Resnikoff, Proceedings of the International Conference on Quality Control, Tokyo, Japan D3, 7-11.

MIL-STD-414 (1957). Sampling procedures and tables for inspection by variables for percent defectives, MIL-STD-414, U.S. Government Printing Office, Washington, D.C.

Nelson, W. and Schmee, J. (1979). Inference for (log) normal life distributions from small singly censored samples and BLUE's, Technometrics 21, 43-54.

Owen, D. B. (1969). Summary of recent work on variables acceptance sampling with emphasis on non-normality, Technometrics 11, 631-637.

Persson, T. and Rootzen, H. (1977). Simple and highly efficient estimators for a Type I censored normal sample, Biometrika 64, 123-128.

Rao, J. N. K., Subrahmaniam, K., and Owen, D. B. (1972). Effect of non-normality on tolerance limits which control percentages in both tails of normal distribution, Technometrics 14, 571-575.

Schneider, H. and Wilrich, P. Th. (1981). The robustness of sampling plans for inspection by variables, in Computational Statistics, H. Büning and P. Naeve, eds., Walter de Gruyter, Berlin, New York.

Schneider, H. (1985). The performance of variable sampling plans when the normal distribution is truncated, Journal of Quality Technology 17, 74-80.

Srivastava, A. B. L. (1961). Variables sampling inspection for non-normal samples, Journal of Science Engineering Research 5, 145-152.

Takagi, K. (1972). On designing unknown-sigma sampling plans based on a wide class of non-normal distributions, Technometrics 14, 669-678.

Tiku, M. L. (1967). Estimating the mean and standard deviation from a censored sample, Biometrika 54, 155-165.

Tiku, M. L. (1980). Robustness of MML estimators based on censored samples and robust test statistics, Journal of Statistical Planning and Inference 4, 123-143.

Wallis, W. A. (1947). Techniques of Statistical Analysis, McGraw-Hill, New York, p. 17.

Wilrich, P. Th. (1970). Nomogramme zur Ermittlung von Stichprobenplänen fur messende Prüfung bei einseitig vorgeschriebenen Toleranzgrenzen, Qualität und Zuverlässigkeit 15, 61-65 and 16, 181-187.

APPENDIX A

Nomographs and Tables

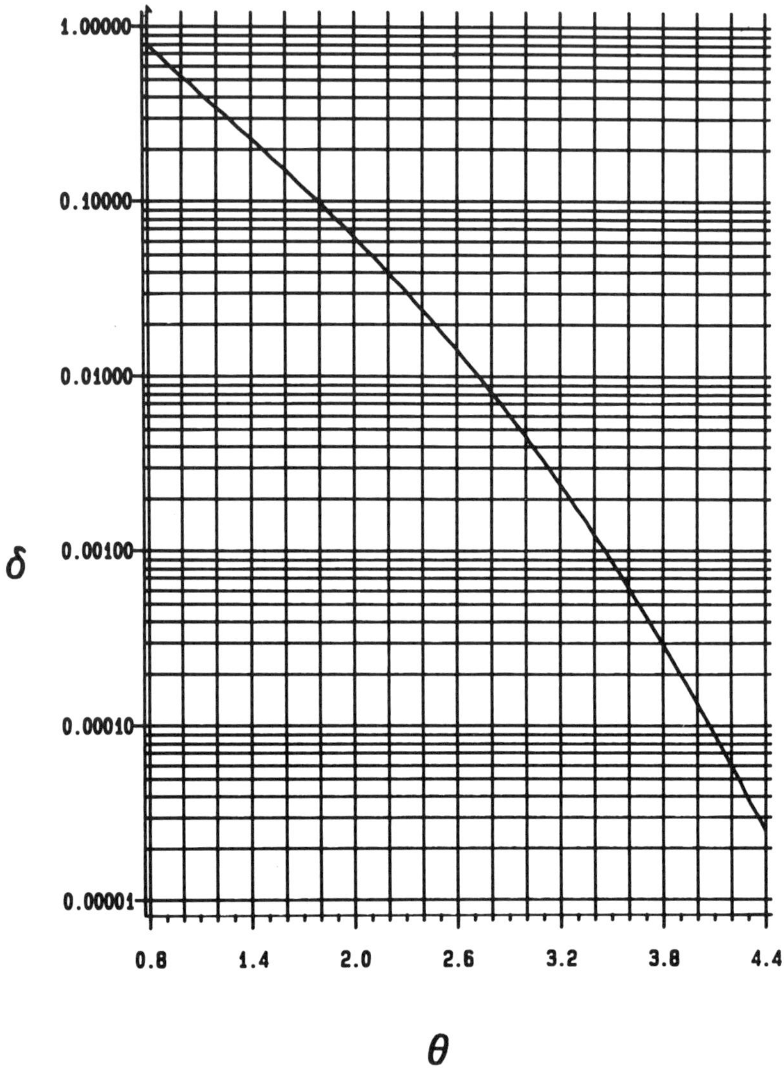

FIGURE A1 Nomograph for singly truncated normal distribution with known σ.

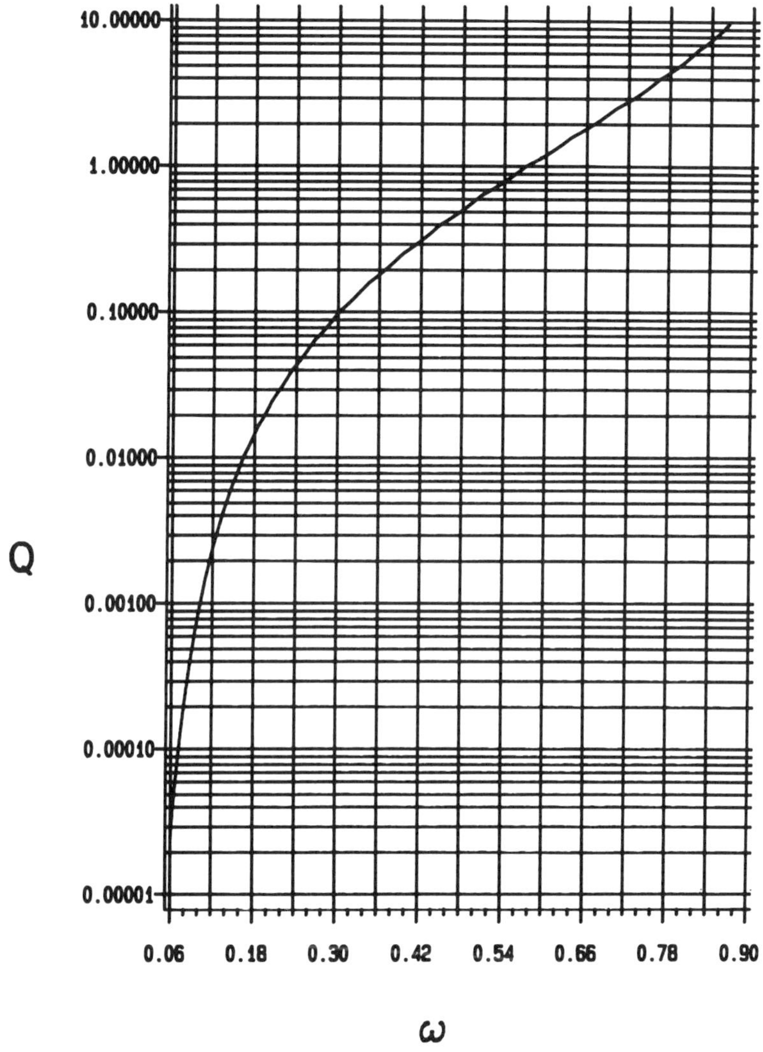

FIGURE A2 Nomograph for singly truncated normal distribution with unknown σ.

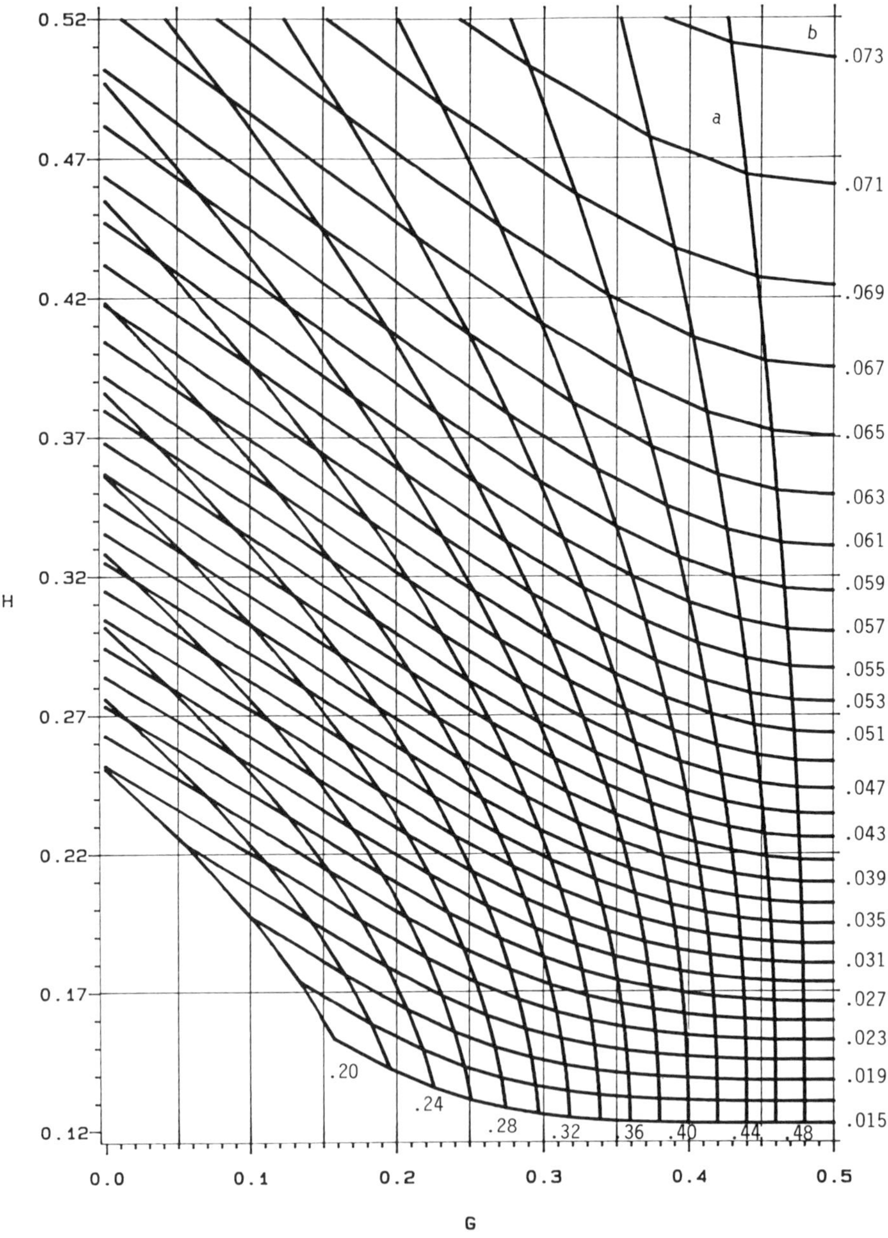

FIGURE A3 Nomograph for doubly truncated normal distribution with unknown σ.

TABLE A1 The Function Q(w) for Estimating the Mean and Standard Deviation from a Singly Truncated Sample with Unknown σ

w	.0000	.0005	.0010	.0015	.0020	.0025
0.060	.000013	.000016	.000019	.000023	.000026	.000029
0.063	.000033	.000036	.000040	.000043	.000047	.000051
0.066	.000055	.000058	.000062	.000067	.000071	.000075
0.069	.000080	.000084	.000089	.000094	.000099	.000104
0.072	.000110	.000115	.000121	.000127	.000133	.000140
0.075	.000146	.000153	.000160	.000168	.000175	.000183
0.078	.000191	.000200	.000208	.000217	.000226	.000236
0.081	.000246	.000256	.000266	.000277	.000288	.000299
0.084	.000311	.000323	.000335	.000348	.000361	.000375
0.087	.000389	.000403	.000417	.000432	.000448	.000463
0.090	.000480	.000496	.000513	.000531	.000549	.000567
0.093	.000586	.000605	.000624	.000645	.000665	.000686
0.096	.000708	.000730	.000752	.000775	.000799	.000822
0.099	.000847	.000872	.000897	.000923	.000950	.000977
0.102	.001004	.001033	.001061	.001090	.001120	.001151
0.105	.001181	.001213	.001245	.001278	.001311	.001345
0.108	.001379	.001414	.001450	.001486	.001523	.001560
0.111	.001598	.001637	.001676	.001716	.001757	.001798
0.114	.001840	.001883	.001926	.001970	.002014	.002059
0.117	.002105	.002152	.002199	.002247	.002296	.002345
0.120	.002395	.002446	.002498	.002550	.002603	.002657
0.123	.002711	.002766	.002822	.002879	.002936	.002994
0.126	.003053	.003113	.003173	.003234	.003296	.003359
0.129	.003423	.003487	.003552	.003618	.003685	.003752
0.132	.003821	.003890	.003960	.004030	.004102	.004174
0.135	.004248	.004322	.004397	.004473	.004549	.004627
0.138	.004705	.004784	.004864	.004945	.005027	.005110
0.141	.005194	.005278	.005364	.005450	.005537	.005625
0.144	.005714	.005804	.005895	.005987	.006079	.006173
0.147	.006267	.006363	.006459	.006556	.006655	.006754
0.150	.006854	.006955	.007057	.007160	.007265	.007370
0.153	.007476	.007583	.007691	.007799	.007909	.008020
0.156	.008132	.008245	.008359	.008474	.008590	.008707
0.159	.008825	.008944	.009064	.009186	.009308	.009431
0.162	.009555	.009681	.009807	.009934	.010063	.010193
0.165	.010323	.010455	.010588	.010722	.010857	.010993
0.168	.011130	.011268	.011407	.011548	.011689	.011832
0.171	.011976	.012121	.012267	.012414	.012562	.012712
0.174	.012862	.013014	.013167	.013321	.013476	.013633
0.177	.013790	.013949	.014109	.014270	.014432	.014595
0.180	.014760	.014926	.015093	.015261	.015430	.015601
0.183	.015772	.015945	.016120	.016295	.016472	.016650
0.186	.016829	.017009	.017191	.017373	.017558	.017743
0.189	.017929	.018117	.018306	.018497	.018688	.018881
0.192	.019076	.019271	.019468	.019666	.019865	.020066
0.195	.020268	.020471	.020676	.020882	.021089	.021298
0.198	.021507	.021719	.021931	.022145	.022360	.022577

TABLE A1 (continued)

w	.0000	.0005	.0010	.0015	.0020	.0025
0.201	.022795	.023014	.023235	.023457	.023680	.023905
0.204	.024131	.024358	.024587	.024817	.025049	.025282
0.207	.025516	.025752	.025989	.026228	.026468	.026710
0.210	.026953	.027197	.027443	.027690	.027938	.028188
0.213	.028440	.028693	.028947	.029203	.029461	.029719
0.216	.029980	.030241	.030505	.030769	.031036	.031303
0.219	.031572	.031843	.032115	.032389	.032664	.032941
0.222	.033219	.033499	.033780	.034063	.034347	.034633
0.225	.034921	.035210	.035500	.035792	.036086	.036381
0.228	.036678	.036977	.037277	.037578	.037881	.038186
0.231	.038492	.038800	.039110	.039421	.039734	.040048
0.234	.040364	.040682	.041001	.041322	.041645	.041969
0.237	.042295	.042622	.042951	.043282	.043615	.043949
0.240	.044284	.044622	.044961	.045302	.045645	.045989
0.243	.046335	.046683	.047032	.047383	.047736	.048090
0.246	.048447	.048805	.049164	.049526	.049889	.050254
0.249	.050621	.050989	.051360	.051732	.052106	.052481
0.252	.052859	.053238	.053619	.054002	.054386	.054773
0.255	.055161	.055551	.055943	.056336	.056732	.057129
0.258	.057528	.057929	.058332	.058737	.059143	.059552
0.261	.059962	.060374	.060789	.061205	.061622	.062042
0.264	.062464	.062887	.063313	.063740	.064170	.064601
0.267	.065034	.065469	.065906	.066345	.066786	.067229
0.270	.067674	.068120	.068569	.069020	.069473	.069927
0.273	.070384	.070843	.071304	.071766	.072231	.072698
0.276	.073166	.073637	.074110	.074585	.075062	.075541
0.279	.076022	.076505	.076990	.077477	.077966	.078458
0.282	.078951	.079447	.079944	.080444	.080946	.081450
0.285	.081956	.082464	.082974	.083487	.084001	.084518
0.288	.085037	.085558	.086081	.086606	.087134	.087664
0.291	.088195	.088729	.089266	.089804	.090345	.090888
0.294	.091433	.091980	.092530	.093081	.093636	.094192
0.297	.094750	.095311	.095874	.096439	.097007	.097577
0.300	.098149	.098724	.099300	.099879	.100461	.101044
0.303	.101630	.102219	.102809	.103402	.103998	.104596
0.306	.105196	.105798	.106403	.107010	.107620	.108232
0.309	.108846	.109463	.110082	.110703	.111327	.111954
0.312	.112583	.113214	.113848	.114484	.115123	.115764
0.315	.116407	.117053	.117702	.118353	.119007	.119663
0.318	.120321	.120982	.121646	.122312	.122981	.123652
0.321	.124326	.125002	.125681	.126362	.127046	.127733
0.324	.128422	.129114	.129808	.130505	.131205	.131907
0.327	.132612	.133319	.134030	.134742	.135458	.136176
0.330	.136897	.137620	.138346	.139075	.139807	.140541
0.333	.141278	.142018	.142760	.143505	.144253	.145004
0.336	.145757	.146513	.147272	.148034	.148799	.149566
0.339	.150336	.151109	.151885	.152663	.153445	.154229
0.342	.155016	.155806	.156599	.157394	.158193	.158994
0.345	.159798	.160606	.161416	.162229	.163045	.163864

TABLE A1 (continued)

w	.0000	.0005	.0010	.0015	.0020	.0025
0.348	.164685	.165510	.166338	.167168	.168002	.168839
0.351	.169678	.170521	.171366	.172215	.173067	.173921
0.354	.174779	.175640	.176503	.177370	.178240	.179113
0.357	.179989	.180868	.181750	.182636	.183524	.184416
0.360	.185310	.186208	.187109	.188013	.188920	.189831
0.363	.190745	.191662	.192582	.193505	.194431	.195361
0.366	.196294	.197230	.198170	.199112	.200058	.201007
0.369	.201960	.202916	.203875	.204837	.205803	.206772
0.372	.207744	.208720	.209699	.210682	.211667	.212657
0.375	.213649	.214645	.215645	.216648	.217654	.218664
0.378	.219677	.220694	.221714	.222737	.223764	.224795
0.381	.225829	.226867	.227908	.228952	.230001	.231053
0.384	.232108	.233167	.234230	.235295	.236365	.237439
0.387	.238515	.239596	.240680	.241768	.242860	.243955
0.390	.245054	.246156	.247263	.248372	.249486	.250604
0.393	.251725	.252850	.253979	.255111	.256248	.257388
0.396	.258532	.259679	.260831	.261986	.263146	.264309
0.399	.265476	.266647	.267822	.269000	.270183	.271370
0.402	.272560	.273754	.274953	.276155	.277361	.278572
0.405	.279786	.281005	.282227	.283454	.284684	.285918
0.408	.287157	.288400	.289647	.290898	.292153	.293412
0.411	.294675	.295943	.297215	.298490	.299770	.301055
0.414	.302343	.303636	.304933	.306234	.307539	.308849
0.417	.310163	.311481	.312804	.314131	.315462	.316798
0.420	.318138	.319482	.320831	.322184	.323541	.324904
0.423	.326270	.327641	.329016	.330396	.331780	.333169
0.426	.334563	.335961	.337363	.338770	.340181	.341598
0.429	.343018	.344443	.345874	.347308	.348747	.350191
0.432	.351640	.353093	.354551	.356014	.357481	.358953
0.435	.360430	.361912	.363398	.364890	.366386	.367887
0.438	.369393	.370903	.372418	.373939	.375464	.376994
0.441	.378529	.380070	.381615	.383165	.384720	.386280
0.444	.387845	.389415	.390990	.392570	.394156	.395746
0.447	.397341	.398941	.400547	.402158	.403774	.405395
0.450	.407022	.408653	.410290	.411932	.413580	.415232
0.453	.416891	.418553	.420222	.421896	.423576	.425260
0.456	.426950	.428646	.430346	.432053	.433765	.435482
0.459	.437205	.438933	.440667	.442406	.444151	.445901
0.462	.447657	.449419	.451186	.452959	.454737	.456522
0.465	.458312	.460108	.461909	.463716	.465529	.467348
0.468	.469172	.471003	.472839	.474681	.476529	.478382
0.471	.480242	.482107	.483979	.485856	.487740	.489630
0.474	.491525	.493427	.495334	.497247	.499168	.501093
0.477	.503026	.504964	.506908	.508859	.510815	.512778
0.480	.514747	.516723	.518704	.520692	.522687	.524688
0.483	.526695	.528708	.530728	.532754	.534787	.536827
0.486	.538872	.540924	.542983	.545048	.547120	.549199
0.489	.551284	.553375	.555474	.557579	.559690	.561809
0.492	.563934	.566066	.568205	.570350	.572503	.574662
0.495	.576828	.579001	.581181	.583368	.585562	.587762

TABLE A1 (continued)

w	.0000	.0005	.0010	.0015	.0020	.0025
0.498	.589970	.592185	.594407	.596636	.598871	.601115
0.501	.603365	.605623	.607887	.610159	.612438	.614725
0.504	.617017	.619319	.621626	.623942	.626265	.628595
0.507	.630934	.633279	.635631	.637991	.640359	.642734
0.510	.645117	.647508	.649906	.652311	.654725	.657146
0.513	.659574	.662011	.664455	.666907	.669367	.671835
0.516	.674311	.676794	.679286	.681785	.684292	.686808
0.519	.689332	.691863	.694402	.696950	.699506	.702071
0.522	.704643	.707222	.709811	.712408	.715014	.717627
0.525	.720250	.722880	.725519	.728166	.730822	.733486
0.528	.736159	.738840	.741530	.744228	.746936	.749652
0.531	.752376	.755109	.757851	.760603	.763362	.766131
0.534	.768909	.771694	.774490	.777294	.780108	.782930
0.537	.785762	.788602	.791452	.794311	.797179	.800056
0.540	.802943	.805839	.808744	.811657	.814582	.817515
0.543	.820458	.823411	.826372	.829343	.832324	.835315
0.546	.838316	.841325	.844345	.847374	.850414	.853464
0.549	.856522	.859590	.862670	.865759	.868858	.871965
0.552	.875085	.878213	.881352	.884501	.887661	.890831
0.555	.894011	.897201	.900402	.903613	.906835	.910067
0.558	.913309	.916562	.919827	.923101	.926386	.929682
0.561	.932988	.936306	.939633	.942972	.946322	.949683
0.564	.953055	.956437	.959830	.963235	.966651	.970080
0.567	.973518	.976967	.980428	.983899	.987383	.990879
0.570	.994386	.997904	.273221	.273754	.274288	.274821
0.573	.275356	.275892	.276428	.276965	.277503	.278042
0.576	.278581	.279121	.279662	.280203	.280746	.281289
0.579	.281833	.282377	.282922	.283468	.284015	.284563
0.582	.285111	.285660	.286209	.286760	.287311	.287863
0.585	.288416	.288969	.289523	.290078	.290634	.291190
0.588	.291747	.292305	.292863	.293423	.293982	.294543
0.591	.295104	.295666	.296229	.296793	.297357	.297922
0.594	.298488	.299055	.299622	.300189	.300758	.301328
0.597	.301898	.302468	.303040	.303612	.304185	.304758
0.600	.305333	.305908	.306484	.307060	.307637	.308215
0.603	.308794	.309373	.309953	.310534	.311116	.311698
0.606	.312281	.312864	.313449	.314033	.314619	.315205
0.609	.315792	.316380	.316969	.317558	.318148	.318738
0.612	.319330	.319921	.320514	.321108	.321702	.322297
0.615	.322892	.323488	.324085	.324682	.325281	.325880
0.618	.326479	.327079	.327680	.328282	.328884	.329487
0.621	.330091	.330696	.331301	.331906	.332513	.333120
0.624	.333728	.334336	.334946	.335555	.336166	.336777
0.627	.337389	.338002	.338615	.339229	.339843	.340459
0.630	.341075	.341691	.342309	.342927	.343546	.344165
0.633	.344785	.345405	.346027	.346649	.347271	.347895
0.636	.348519	.349144	.349769	.350395	.351022	.351649
0.639	.352277	.352906	.353535	.354165	.354796	.355427
0.642	.356059	.356692	.357325	.357959	.358593	.359229
0.645	.359865	.360501	.361139	.361776	.362415	.363054

TABLE A2 Auxiliary Estimation Function, λ, for Singly Censored Samples

ω \ h=	.01	.02	.03	.04	.05	.06	.07	.08	.09	.10
0.01	.01020	.02059	.03118	.04197	.05297	.06417	.07557	.08719	.09902	.1111
0.02	.01029	.02077	.03145	.04233	.05341	.06469	.07618	.08787	.09978	.1119
0.03	.01038	.02095	.03172	.04268	.05384	.06520	.07677	.08854	.10052	.1127
0.04	.01047	.02113	.03197	.04302	.05426	.06570	.07734	.08919	.10125	.1135
0.05	.01055	.02129	.03223	.04335	.05467	.06619	.07791	.08983	.10197	.1143
0.06	.01064	.02146	.03247	.04367	.05507	.06667	.07846	.09046	.10267	.1151
0.07	.01072	.02162	.03271	.04399	.05546	.06713	.07900	.09107	.10335	.1158
0.08	.01080	.02178	.03294	.04430	.05585	.06759	.07953	.09168	.10403	.1166
0.09	.01087	.02193	.03317	.04460	.05623	.06804	.08006	.09227	.10469	.1173
0.10	.01095	.02208	.03340	.04490	.05660	.06848	.08057	.09285	.10534	.1180
0.11	.01102	.02223	.03362	.04519	.05696	.06892	.08107	.09343	.10598	.1187
0.12	.01110	.02238	.03384	.04548	.05732	.06934	.08157	.09399	.10661	.1194
0.13	.01117	.02252	.03405	.04577	.05767	.06976	.08205	.09454	.10723	.1201
0.14	.01124	.02266	.03426	.04604	.05802	.07018	.08254	.09509	.10785	.1208
0.15	.01131	.02280	.03447	.04632	.05836	.07059	.08301	.09563	.10845	.1215
0.16	.01138	.02293	.03467	.04659	.05869	.07099	.08348	.09616	.10905	.1221
0.17	.01145	.02307	.03487	.04685	.05902	.07138	.08394	.09668	.10963	.1228
0.18	.01151	.02320	.03507	.04712	.05935	.07177	.08439	.09720	.11021	.1234
0.19	.01158	.02333	.03526	.04737	.05967	.07216	.08484	.09771	.11079	.1241
0.20	.01164	.02346	.03545	.04763	.05999	.07254	.08528	.09822	.11135	.1247
0.21	.01171	.02359	.03564	.04788	.06030	.07291	.08572	.09871	.11191	.1253
0.22	.01177	.02371	.03583	.04813	.06061	.07329	.08615	.09921	.11246	.1259
0.23	.01183	.02383	.03601	.04838	.06092	.07365	.08657	.09969	.11301	.1265
0.24	.01189	.02396	.03620	.04862	.06122	.07401	.08700	.10017	.11355	.1271
0.25	.01195	.02408	.03638	.04886	.06152	.07437	.08741	.10065	.11408	.1277
0.26	.01201	.02420	.03656	.04909	.06182	.07473	.08783	.10112	.11461	.1283
0.27	.01207	.02431	.03673	.04933	.06211	.07508	.08823	.10158	.11513	.1289
0.28	.01213	.02443	.03691	.04956	.06240	.07542	.08864	.10205	.11565	.1295
0.29	.01219	.02454	.03708	.04979	.06269	.07577	.08904	.10250	.11616	.1300
0.30	.01224	.02466	.03725	.05002	.06297	.07611	.08943	.10295	.11667	.1306
0.31	.01230	.02477	.03742	.05024	.06325	.07644	.08982	.10340	.11717	.1311
0.32	.01236	.02488	.03758	.05047	.06353	.07678	.09021	.10384	.11767	.1317
0.33	.01241	.02499	.03775	.05069	.06380	.07711	.09060	.10428	.11816	.1322
0.34	.01247	.02510	.03791	.05090	.06408	.07743	.09098	.10472	.11865	.1328
0.35	.01252	.02521	.03808	.05112	.06435	.07776	.09136	.10515	.11914	.1333
0.36	.01257	.02532	.03824	.05133	.06461	.07808	.09173	.10557	.11962	.1339
0.37	.01263	.02542	.03840	.05155	.06488	.07839	.09210	.10600	.12009	.1344
0.38	.01268	.02553	.03855	.05176	.06514	.07871	.09247	.10642	.12057	.1349
0.39	.01273	.02563	.03871	.05197	.06540	.07902	.09283	.10683	.12103	.1354
0.40	.01278	.02574	.03887	.05217	.06566	.07933	.09319	.10725	.12150	.1360
0.41	.01284	.02584	.03902	.05238	.06592	.07964	.09355	.10766	.12196	.1365
0.42	.01289	.02594	.03917	.05258	.06617	.07994	.09391	.10806	.12242	.1370
0.43	.01294	.02604	.03932	.05278	.06642	.08025	.09426	.10847	.12287	.1375
0.44	.01299	.02614	.03947	.05298	.06667	.08055	.09461	.10887	.12332	.1380
0.45	.01304	.02624	.03962	.05318	.06692	.08084	.09496	.10926	.12377	.1385
0.46	.01309	.02634	.03977	.05338	.06717	.08114	.09530	.10966	.12421	.1390
0.47	.01313	.02644	.03992	.05357	.06741	.08143	.09565	.11005	.12465	.1395
0.48	.01318	.02654	.04006	.05377	.06765	.08173	.09598	.11044	.12509	.1399
0.49	.01323	.02663	.04021	.05396	.06790	.08201	.09632	.11082	.12552	.1404
0.50	.01328	.02673	.04035	.05415	.06813	.08230	.09666	.11121	.12595	.1409

TABLE A2 (continued)

ω	h= .01	.02	.03	.04	.05	.06	.07	.08	.09	.10
0.52	.01337	.02692	.04064	.05453	.06861	.08287	.09732	.11196	.12681	.1418
0.54	.01347	.02710	.04092	.05490	.06907	.08343	.09797	.11271	.12765	.1428
0.56	.01356	.02729	.04119	.05527	.06954	.08398	.09862	.11345	.12848	.1437
0.58	.01365	.02747	.04146	.05564	.06999	.08453	.09926	.11418	.12930	.1446
0.60	.01374	.02765	.04173	.05600	.07044	.08507	.09989	.11490	.13011	.1455
0.62	.01383	.02783	.04200	.05635	.07088	.08560	.10051	.11561	.13091	.1464
0.64	.01391	.02800	.04226	.05670	.07132	.08613	.10112	.11631	.13170	.1473
0.66	.01400	.02817	.04252	.05705	.07175	.08665	.10173	.11701	.13248	.1482
0.68	.01409	.02834	.04278	.05739	.07218	.08716	.10233	.11769	.13326	.1490
0.70	.01417	.02851	.04303	.05773	.07260	.08767	.10292	.11837	.13402	.1499
0.72	.01425	.02868	.04328	.05806	.07302	.08817	.10351	.11905	.13478	.1507
0.74	.01434	.02885	.04353	.05839	.07344	.08867	.10409	.11971	.13553	.1515
0.76	.01442	.02901	.04377	.05872	.07385	.08916	.10467	.12037	.13627	.1524
0.78	.01450	.02917	.04402	.05904	.07425	.08965	.10524	.12102	.13700	.1532
0.80	.01458	.02933	.04426	.05936	.07465	.09013	.10580	.12167	.13773	.1540
0.82	.01466	.02949	.04450	.05968	.07505	.09061	.10636	.12231	.13845	.1548
0.84	.01474	.02965	.04473	.06000	.07545	.09108	.10691	.12294	.13916	.1556
0.86	.01481	.02980	.04496	.06031	.07584	.09155	.10746	.12357	.13987	.1564
0.88	.01489	.02995	.04520	.06062	.07622	.09202	.10800	.12419	.14057	.1572
0.90	.01497	.03011	.04542	.06092	.07661	.09248	.10854	.12480	.14126	.1579
0.92	.01504	.03026	.04565	.06123	.07699	.09293	.10907	.12541	.14195	.1587
0.94	.01512	.03041	.04588	.06153	.07736	.09339	.10960	.12602	.14263	.1595
0.96	.01519	.03056	.04610	.06182	.07773	.09383	.11013	.12662	.14331	.1602
0.98	.01526	.03070	.04632	.06212	.07810	.09428	.11065	.12721	.14398	.1610
1.00	.01534	.03085	.04654	.06241	.07847	.09472	.11116	.12780	.14465	.1617
1.02	.01541	.03099	.04676	.06270	.07884	.09516	.11167	.12839	.14531	.1624
1.04	.01548	.03114	.04697	.06299	.07920	.09559	.11218	.12897	.14596	.1632
1.06	.01555	.03128	.04719	.06328	.07955	.09602	.11268	.12954	.14661	.1639
1.08	.01562	.03142	.04740	.06356	.07991	.09645	.11318	.13012	.14725	.1646
1.10	.01569	.03156	.04761	.06384	.08026	.09687	.11368	.13068	.14789	.1653
1.12	.01576	.03170	.04782	.06412	.08061	.09729	.11417	.13125	.14853	.1660
1.14	.01583	.03184	.04803	.06440	.08096	.09771	.11466	.13181	.14916	.1667
1.16	.01590	.03198	.04823	.06467	.08130	.09812	.11514	.13236	.14978	.1674
1.18	.01597	.03211	.04844	.06495	.08165	.09854	.11562	.13291	.15041	.1681
1.20	.01604	.03225	.04864	.06522	.08199	.09895	.11610	.13346	.15102	.1688
1.22	.01610	.03238	.04884	.06549	.08232	.09935	.11658	.13400	.15163	.1695
1.24	.01617	.03251	.04904	.06576	.08266	.09975	.11705	.13454	.15224	.1702
1.26	.01623	.03265	.04924	.06602	.08299	.10015	.11751	.13508	.15285	.1708
1.28	.01630	.03278	.04944	.06629	.08332	.10055	.11798	.13561	.15345	.1715
1.30	.01637	.03291	.04963	.06655	.08365	.10095	.11844	.13614	.15404	.1722
1.32	.01643	.03304	.04983	.06681	.08398	.10134	.11890	.13666	.15464	.1728
1.34	.01649	.03317	.05002	.06707	.08430	.10173	.11936	.13719	.15523	.1735
1.36	.01656	.03330	.05022	.06732	.08462	.10212	.11981	.13771	.15581	.1741
1.38	.01662	.03342	.05041	.06758	.08494	.10250	.12026	.13822	.15639	.1748
1.40	.01668	.03355	.05060	.06783	.08526	.10288	.12071	.13873	.15697	.1754
1.42	.01675	.03367	.05079	.06808	.08558	.10326	.12115	.13924	.15754	.1761
1.44	.01681	.03380	.05097	.06834	.08589	.10364	.12159	.13975	.15811	.1767
1.46	.01687	.03392	.05116	.06858	.08620	.10402	.12203	.14025	.15868	.1773
1.48	.01693	.03405	.05135	.06883	.08651	.10439	.12247	.14075	.15925	.1780

TABLE A2 (continued)

ω	h= 0.15	0.20	0.25	0.30	0.35	0.40	0.45	0.50
0.01	0.1747	0.2443	0.3205	0.4043	0.4967	0.5989	0.7128	0.8403
0.04	0.1782	0.2488	0.3261	0.4109	0.5042	0.6074	0.7221	0.8506
0.07	0.1816	0.2532	0.3315	0.4172	0.5114	0.6155	0.7312	0.8606
0.10	0.1848	0.2574	0.3366	0.4233	0.5184	0.6234	0.7400	0.8703
0.13	0.1879	0.2614	0.3416	0.4291	0.5252	0.6311	0.7486	0.8798
0.16	0.1908	0.2653	0.3464	0.4348	0.5318	0.6386	0.7570	0.8891
0.19	0.1937	0.2691	0.3510	0.4404	0.5382	0.6459	0.7652	0.8982
0.22	0.1964	0.2727	0.3555	0.4458	0.5445	0.6530	0.7732	0.9071
0.25	0.1991	0.2763	0.3599	0.4510	0.5506	0.6600	0.7810	0.9158
0.28	0.2017	0.2797	0.3642	0.4561	0.5565	0.6668	0.7887	0.9243
0.31	0.2042	0.2830	0.3684	0.4611	0.5624	0.6735	0.7962	0.9327
0.34	0.2067	0.2863	0.3725	0.4660	0.5681	0.6800	0.8035	0.9410
0.37	0.2091	0.2895	0.3764	0.4708	0.5736	0.6864	0.8108	0.9490
0.40	0.2114	0.2926	0.3803	0.4755	0.5791	0.6927	0.8179	0.9570
0.43	0.2137	0.2957	0.3841	0.4801	0.5845	0.6989	0.8249	0.9648
0.46	0.2159	0.2986	0.3879	0.4846	0.5898	0.7049	0.8318	0.9725
0.49	0.2181	0.3016	0.3916	0.4890	0.5950	0.7109	0.8385	0.9801
0.52	0.2202	0.3044	0.3952	0.4933	0.6001	0.7168	0.8452	0.9876
0.55	0.2224	0.3073	0.3987	0.4976	0.6051	0.7225	0.8517	0.9950
0.58	0.2244	0.3100	0.4022	0.5018	0.6100	0.7282	0.8582	1.0022
0.61	0.2265	0.3127	0.4056	0.5059	0.6149	0.7338	0.8646	1.0094
0.64	0.2284	0.3154	0.4090	0.5100	0.6197	0.7394	0.8709	1.0165
0.67	0.2304	0.3181	0.4123	0.5140	0.6244	0.7448	0.8771	1.0234
0.70	0.2323	0.3206	0.4156	0.5180	0.6291	0.7502	0.8832	1.0303
0.73	0.2342	0.3232	0.4188	0.5219	0.6337	0.7555	0.8892	1.0371
0.76	0.2361	0.3257	0.4219	0.5257	0.6382	0.7608	0.8952	1.0439
0.79	0.2380	0.3282	0.4251	0.5295	0.6427	0.7659	0.9011	1.0505
0.82	0.2398	0.3307	0.4282	0.5333	0.6471	0.7710	0.9069	1.0571
0.85	0.2416	0.3331	0.4312	0.5370	0.6515	0.7761	0.9127	1.0636
0.88	0.2434	0.3355	0.4342	0.5406	0.6558	0.7811	0.9184	1.0700
0.91	0.2451	0.3378	0.4372	0.5442	0.6600	0.7860	0.9240	1.0764
0.94	0.2468	0.3401	0.4401	0.5478	0.6642	0.7909	0.9296	1.0827
0.97	0.2485	0.3424	0.4431	0.5513	0.6684	0.7957	0.9351	1.0889
1.00	0.2502	0.3447	0.4459	0.5548	0.6725	0.8005	0.9406	1.0951
1.03	0.2519	0.3470	0.4488	0.5583	0.6766	0.8052	0.9460	1.1012
1.06	0.2535	0.3492	0.4516	0.5617	0.6806	0.8099	0.9513	1.1072
1.09	0.2552	0.3514	0.4543	0.5650	0.6846	0.8146	0.9566	1.1132
1.12	0.2568	0.3535	0.4571	0.5684	0.6886	0.8191	0.9619	1.1191
1.15	0.2584	0.3557	0.4598	0.5717	0.6925	0.8237	0.9671	1.1250
1.18	0.2599	0.3578	0.4625	0.5750	0.6964	0.8282	0.9722	1.1308
1.21	0.2615	0.3599	0.4652	0.5782	0.7002	0.8327	0.9773	1.1366
1.24	0.2631	0.3620	0.4678	0.5814	0.7040	0.8371	0.9824	1.1424
1.27	0.2646	0.3641	0.4704	0.5846	0.7078	0.8415	0.9874	1.1480
1.30	0.2661	0.3661	0.4730	0.5878	0.7115	0.8458	0.9924	1.1537
1.33	0.2676	0.3681	0.4756	0.5909	0.7152	0.8501	0.9973	1.1593
1.36	0.2691	0.3702	0.4781	0.5940	0.7189	0.8544	1.0022	1.1648
1.39	0.2706	0.3721	0.4807	0.5971	0.7226	0.8586	1.0070	1.1703
1.42	0.2720	0.3741	0.4832	0.6001	0.7262	0.8628	1.0119	1.1757
1.45	0.2735	0.3761	0.4856	0.6031	0.7298	0.8670	1.0166	1.1812
1.48	0.2749	0.3780	0.4881	0.6061	0.7333	0.8711	1.0214	1.1865

TABLE A2 (continued)

ω	h= 0.55	0.60	0.65	0.70	0.75	0.80	0.85	0.90
0.01	0.9846	1.1495	1.3406	1.5662	1.8393	2.1814	2.6337	3.2890
0.04	0.9958	1.1616	1.3537	1.5803	1.8544	2.1977	2.6512	3.3079
0.07	1.0067	1.1735	1.3665	1.5942	1.8693	2.2137	2.6684	3.3266
0.10	1.0174	1.1851	1.3791	1.6078	1.8840	2.2295	2.6855	3.3452
0.13	1.0278	1.1965	1.3915	1.6211	1.8984	2.2450	2.7023	3.3635
0.16	1.0380	1.2076	1.4036	1.6342	1.9126	2.2604	2.7189	3.3816
0.19	1.0480	1.2185	1.4155	1.6471	1.9265	2.2755	2.7353	3.3995
0.22	1.0578	1.2293	1.4272	1.6598	1.9403	2.2904	2.7515	3.4172
0.25	1.0674	1.2398	1.4387	1.6724	1.9539	2.3052	2.7676	3.4347
0.28	1.0768	1.2502	1.4500	1.6847	1.9673	2.3197	2.7834	3.4521
0.31	1.0861	1.2603	1.4611	1.6968	1.9805	2.3341	2.7991	3.4693
0.34	1.0952	1.2704	1.4721	1.7088	1.9935	2.3483	2.8146	3.4863
0.37	1.1042	1.2802	1.4829	1.7206	2.0064	2.3623	2.8299	3.5032
0.40	1.1130	1.2900	1.4936	1.7322	2.0191	2.3762	2.8451	3.5199
0.43	1.1217	1.2995	1.5041	1.7437	2.0316	2.3899	2.8601	3.5365
0.46	1.1302	1.3090	1.5144	1.7551	2.0441	2.4034	2.8750	3.5529
0.49	1.1387	1.3183	1.5247	1.7663	2.0563	2.4169	2.8897	3.5692
0.52	1.1470	1.3275	1.5348	1.7774	2.0684	2.4301	2.9042	3.5853
0.55	1.1552	1.3365	1.5448	1.7883	2.0804	2.4433	2.9187	3.6013
0.58	1.1633	1.3455	1.5546	1.7992	2.0923	2.4563	2.9330	3.6171
0.61	1.1713	1.3543	1.5644	1.8099	2.1040	2.4692	2.9472	3.6329
0.64	1.1791	1.3631	1.5740	1.8205	2.1157	2.4819	2.9612	3.6485
0.67	1.1869	1.3717	1.5835	1.8309	2.1272	2.4945	2.9751	3.6639
0.70	1.1946	1.3802	1.5929	1.8413	2.1385	2.5071	2.9889	3.6793
0.73	1.2022	1.3887	1.6023	1.8515	2.1498	2.5195	3.0026	3.6945
0.76	1.2097	1.3970	1.6115	1.8617	2.1610	2.5317	3.0162	3.7097
0.79	1.2172	1.4053	1.6206	1.8717	2.1720	2.5439	3.0296	3.7247
0.82	1.2245	1.4134	1.6296	1.8817	2.1830	2.5560	3.0430	3.7396
0.85	1.2318	1.4215	1.6386	1.8916	2.1939	2.5680	3.0562	3.7543
0.88	1.2390	1.4295	1.6474	1.9013	2.2046	2.5798	3.0694	3.7690
0.91	1.2461	1.4374	1.6562	1.9110	2.2153	2.5916	3.0824	3.7836
0.94	1.2532	1.4453	1.6649	1.9206	2.2259	2.6033	3.0954	3.7981
0.97	1.2601	1.4530	1.6735	1.9301	2.2364	2.6149	3.1082	3.8125
1.00	1.2670	1.4607	1.6820	1.9395	2.2468	2.6264	3.1210	3.8267
1.03	1.2739	1.4683	1.6904	1.9489	2.2571	2.6378	3.1336	3.8409
1.06	1.2807	1.4759	1.6988	1.9581	2.2673	2.6491	3.1462	3.8550
1.09	1.2874	1.4834	1.7071	1.9673	2.2775	2.6603	3.1587	3.8690
1.12	1.2940	1.4908	1.7154	1.9764	2.2875	2.6715	3.1710	3.8829
1.15	1.3006	1.4982	1.7235	1.9855	2.2975	2.6825	3.1834	3.8968
1.18	1.3072	1.5055	1.7316	1.9944	2.3075	2.6935	3.1956	3.9105
1.21	1.3137	1.5127	1.7397	2.0033	2.3173	2.7044	3.2077	3.9241
1.24	1.3201	1.5199	1.7476	2.0122	2.3271	2.7153	3.2198	3.9377
1.27	1.3265	1.5270	1.7556	2.0209	2.3368	2.7260	3.2318	3.9512
1.30	1.3328	1.5341	1.7634	2.0296	2.3464	2.7367	3.2437	3.9646
1.33	1.3391	1.5411	1.7712	2.0383	2.3560	2.7473	3.2555	3.9779
1.36	1.3453	1.5480	1.7789	2.0469	2.3655	2.7579	3.2673	3.9912
1.39	1.3515	1.5549	1.7866	2.0554	2.3749	2.7684	3.2789	4.0043
1.42	1.3576	1.5618	1.7942	2.0638	2.3843	2.7788	3.2905	4.0174
1.45	1.3637	1.5686	1.8018	2.0722	2.3936	2.7891	3.3021	4.0305
1.48	1.3697	1.5753	1.8093	2.0806	2.4029	2.7994	3.3136	4.0434

TABLE A3 Variance Factors for Singly Truncated and Singly Censored Samples

u_r	Truncated Samples γ_{11}	Truncated Samples γ_{22}	Truncated Samples γ_{12}	Censored Samples γ_{11}	Censored Samples γ_{22}	Censored Samples γ_{12}	h%	u_r
2.33	1.08981	0.64018	0.11913	1.00098	0.50843	0.00287	1	2.33
2.05	1.17815	0.72298	0.21204	1.00230	0.51669	0.00616	2	2.05
1.88	1.27286	0.79385	0.30124	1.00384	0.52496	0.00974	3	1.88
1.75	1.37491	0.85878	0.38988	1.00558	0.53330	0.01355	4	1.75
1.64	1.48470	0.92014	0.47923	1.00752	0.54174	0.01759	5	1.64
1.55	1.60255	0.97919	0.56996	1.00965	0.55029	0.02185	6	1.55
1.48	1.72871	1.03668	0.66250	1.01196	0.55897	0.02634	7	1.48
1.41	1.86344	1.09310	0.75716	1.01447	0.56780	0.03104	8	1.41
1.34	2.00700	1.14881	0.85415	1.01718	0.57678	0.03598	9	1.34
1.28	2.15968	1.20406	0.95366	1.02009	0.58592	0.04113	10	1.28
1.23	2.32177	1.25906	1.05586	1.02321	0.59525	0.04653	11	1.23
1.17	2.49359	1.31396	1.16090	1.02655	0.60476	0.05216	12	1.17
1.13	2.67547	1.36890	1.26889	1.03012	0.61446	0.05805	13	1.13
1.08	2.86776	1.42399	1.37999	1.03392	0.62437	0.06418	14	1.08
1.04	3.07082	1.47932	1.49430	1.03796	0.63449	0.07058	15	1.04
0.99	3.28505	1.53497	1.61196	1.04227	0.64484	0.07726	16	0.99
0.95	3.51086	1.59102	1.73309	1.04685	0.65543	0.08422	17	0.95
0.92	3.74870	1.64755	1.85783	1.05170	0.66625	0.09147	18	0.92
0.88	3.99901	1.70459	1.98628	1.05686	0.67734	0.09903	19	0.88
0.84	4.26228	1.76223	2.11860	1.06232	0.68869	0.10690	20	0.84
0.81	4.53902	1.82051	2.25491	1.06812	0.70032	0.11511	21	0.81
0.77	4.82978	1.87949	2.39537	1.07425	0.71225	0.12367	22	0.77
0.74	5.13510	1.93922	2.54011	1.08075	0.72447	0.13258	23	0.74
0.71	5.45560	1.99975	2.68929	1.08764	0.73701	0.14187	24	0.71
0.67	5.79189	2.06111	2.84306	1.09492	0.74989	0.15156	25	0.67
0.64	6.14466	2.12338	3.00160	1.10264	0.76311	0.16166	26	0.64
0.61	6.51459	2.18659	3.16506	1.11080	0.77668	0.17218	27	0.61
0.58	6.90244	2.25079	3.33365	1.11944	0.79064	0.18316	28	0.58
0.55	7.30896	2.31603	3.50753	1.12858	0.80499	0.19462	29	0.55
0.52	7.73501	2.38237	3.68693	1.13826	0.81975	0.20657	30	0.52
0.50	8.18142	2.44984	3.87202	1.14850	0.83494	0.21904	31	0.50
0.47	8.64917	2.51852	4.06305	1.15934	0.85058	0.23206	32	0.47
0.44	9.13922	2.58844	4.26024	1.17081	0.86669	0.24566	33	0.44
0.41	9.65255	2.65966	4.46382	1.18296	0.88330	0.25986	34	0.41
0.39	10.19033	2.73225	4.67406	1.19583	0.90043	0.27471	35	0.39
0.36	10.75364	2.80625	4.89121	1.20946	0.91809	0.29022	36	0.36
0.33	11.34384	2.88174	5.11559	1.22390	0.93633	0.30645	37	0.33
0.31	11.96210	2.95876	5.34745	1.23921	0.95517	0.32343	38	0.31
0.28	12.60993	3.03741	5.58715	1.25544	0.97463	0.34121	39	0.28
0.25	13.28873	3.11773	5.83499	1.27266	0.99476	0.35982	40	0.25
0.23	14.00012	3.19980	6.09135	1.29092	1.01558	0.37933	41	0.23
0.20	14.74571	3.28369	6.35656	1.31032	1.03714	0.39977	42	0.20
0.18	15.52742	3.36950	6.63108	1.33091	1.05947	0.42122	43	0.18
0.15	16.34709	3.45731	6.91530	1.35280	1.08262	0.44373	44	0.15
0.13	17.20679	3.54719	7.20968	1.37607	1.10662	0.46736	45	0.13
0.10	18.10863	3.63925	7.51466	1.40083	1.13154	0.49220	46	0.10
0.08	19.05507	3.73358	7.83079	1.42718	1.15743	0.51832	47	0.08
0.05	20.04848	3.83027	8.15858	1.45525	1.18433	0.54580	48	0.05

TABLE A3 (continued)

u_r	Truncated Samples γ_{11}	γ_{22}	γ_{12}	Censored Samples γ_{11}	γ_{22}	γ_{12}	h%	u_r
0.03	21.09178	3.92948	8.49867	1.48518	1.21232	0.57474	49	0.03
0.0	22.18762	4.03127	8.85158	1.51709	1.24145	0.60523	50	0.0
-0.03	23.33937	4.13582	9.21809	1.55117	1.27181	0.63739	51	-0.03
-0.05	24.55017	4.24321	9.59881	1.58757	1.30347	0.67134	52	-0.05
-0.08	25.82382	4.35362	9.99456	1.62650	1.33651	0.70721	53	-0.08
-0.10	27.16414	4.46718	10.40613	1.66817	1.37103	0.74513	54	-0.10
-0.13	28.57545	4.58407	10.83443	1.71281	1.40712	0.78527	55	-0.13
-0.15	30.06229	4.70444	11.28037	1.76068	1.44490	0.82780	56	-0.15
-0.18	31.62975	4.82851	11.74502	1.81208	1.48449	0.87291	57	-0.18
-0.20	33.28310	4.95644	12.22942	1.86733	1.52602	0.92081	58	-0.20
-0.23	35.02843	5.08848	12.73484	1.92679	1.56964	0.97173	59	-0.23
-0.25	36.87215	5.22485	13.26256	1.99085	1.61549	1.02593	60	-0.25
-0.28	38.82129	5.36579	13.81396	2.05997	1.66377	1.08370	61	-0.28
-0.31	40.88390	5.51163	14.39072	2.13466	1.71466	1.14535	62	-0.31
-0.33	43.06778	5.66254	14.99424	2.21548	1.76838	1.21124	63	-0.33
-0.36	45.38316	5.81899	15.62669	2.30308	1.82518	1.28178	64	-0.36
-0.39	47.83998	5.98125	16.28992	2.39819	1.88531	1.35740	65	-0.39
-0.41	50.44968	6.14972	16.98616	2.50163	1.94908	1.43862	66	-0.41
-0.44	53.22501	6.32482	17.71786	2.61436	2.01683	1.52602	67	-0.44
-0.47	56.18045	6.50704	18.48788	2.73746	2.08894	1.62023	68	-0.47
-0.50	59.33128	6.69685	19.29900	2.87218	2.16583	1.72200	69	-0.50
-0.52	62.69531	6.89482	20.15463	3.01994	2.24800	1.83219	70	-0.52
-0.55	66.29262	7.10159	21.05856	3.18242	2.33599	1.95176	71	-0.55
-0.58	70.14575	7.31789	22.01503	3.36157	2.43044	2.08184	72	-0.58
-0.61	74.27951	7.54443	23.02852	3.55965	2.53209	2.22373	73	-0.61
-0.64	78.72314	7.78212	24.10445	3.77934	2.64178	2.37897	74	-0.64
-0.67	83.50998	8.03198	25.24898	4.02381	2.76048	2.54931	75	-0.67
-0.71	88.67668	8.29502	26.46855	4.29682	2.88935	2.73688	76	-0.71
-0.74	94.26801	8.57261	27.77136	4.60290	3.02973	2.94416	77	-0.74
-0.77	100.33427	8.86617	29.16632	4.94750	3.18321	3.17414	78	-0.77
-0.81	106.93381	9.17729	30.66365	5.33728	3.35170	3.43041	79	-0.81
-0.84	114.13780	9.50806	32.27602	5.78039	3.53748	3.71733	80	-0.84
-0.88	122.02582	9.86047	34.01692	6.28694	3.74333	4.04024	81	-0.88
-0.92	130.69911	10.23743	35.90413	6.86958	3.97264	4.40575	82	-0.92
-0.95	140.27335	10.64185	37.95708	7.54434	4.22962	4.82216	83	-0.95
-0.99	150.89464	11.07762	40.20068	8.33180	4.51952	5.29994	84	-0.99
-1.04	162.73618	11.54898	42.66353	9.25870	4.84903	5.85259	85	-1.04
-1.08	176.01997	12.06152	45.38251	10.36042	5.22676	6.49767	86	-1.08
-1.13	191.02890	12.62239	48.40440	11.68459	5.66399	7.25856	87	-1.13
-1.17	208.11192	13.23953	51.78484	13.29664	6.17579	8.16687	88	-1.17
-1.23	227.74049	13.92415	55.59964	15.28880	6.78279	9.26650	89	-1.23
-1.28	250.54506	14.69085	59.94881	17.79460	7.51393	10.62003	90	-1.28
-1.34	277.37256	15.55805	64.96339	21.01402	8.41115	12.31955	91	-1.34
-1.41	309.44800	16.55257	70.83221	25.25885	9.53749	14.50608	92	-1.41
-1.48	348.56177	17.71214	77.82625	31.04022	10.99248	17.40630	93	-1.48
-1.55	397.46826	19.09245	86.35417	39.24954	12.94239	21.40707	94	-1.55

TABLE A4 The Coefficients of the Best Linear Estimates of the Mean in Censored Samples up to Sample Size 10 from Normal Population

n	n_1	n_r	x(1)	x(2)	x(3)	x(4)	x(5)	x(6)	x(7)
3	0	1	.0000	1.0000					
4	0	1	.1161	.2408	.6431				
	0	2	-.4056	1.4056					
	1	1		-1.6834	1.6834				
5	0	1	.1252	.1830	.2147	.4771			
	0	2	-.0638	.1498	.9139				
	0	3	-.7411	1.7411					
	1	1		-1.0101	.0000	1.0101			
	1	2		-2.0201	2.0201				
6	0	1	.1183	.1510	.1680	.1828	.3799		
	0	2	.0185	.1226	.1761	.6828			
	0	3	-.2159	.0649	1.1511				
	0	4	-1.0261	2.0261					
	1	1		.3198	.1802	.1802	.3198		
	1	2		.1539	.1781	.6680			
	1	3		-.4578	1.4578				
	2	2			.5000	.5000			
7	0	1	.1088	.1295	.1400	.1487	.1571	.3159	
	0	2	.0465	.1072	.1375	.1626	.5462		
	0	3	-.0738	.0677	.1375	.8686			
	0	4	-.3474	-.0135	1.3609				
	0	5	-1.2733	2.2733					
	1	1		.2718	.1520	.1524	.1520	.2718	
	1	2		.1748	.1432	.1634	.5186		
	1	3		-.0592	.1270	.9321			
	1	4		-.8716	1.8716				
	2	2			.4157	.1686	.4157		
	2	3			.0000	1.0000			
8	0	1	.0997	.1139	.1208	.1265	.1318	.1370	.2704
	0	2	.0569	.0962	.1153	.1309	.1451	.4555	
	0	3	-.0167	.0677	.1084	.1413	.6993		
	0	4	-.1549	.0176	.1001	1.0372			
	0	5	-.4632	-.0855	1.5487				
	0	6	-1.4915	2.4915					
	1	1		.2367	.1315	.1319	.1319	.1315	.2367
	1	2		.1716	.1222	.1338	.1442	.4282	
	1	3		.0431	.1061	.1406	.7102		
	1	4		-.2519	.0741	1.1778			
	1	5		-1.2462	2.2462				
	2	2			.3569	.1431	.1431	.3569	
	2	3			.1742	.1429	.6829		
	2	4			-.4761	1.4761			
	3	3				.5000	.5000		

TABLE A4 (continued)

n	n_l n_r	x(1)	x(2)	x(3)	x(4)	x(5)	x(6)	x(7)	x(8)	x(9)
9	0 1	.0915	.1018	.1067	.1106	.1142	.1177	.1212	.2365	
	0 2	.0602	.0876	.1006	.1110	.1204	.1294	.3909		
	0 3	.0104	.0660	.0923	.1133	.1320	.5860			
	0 4	-.0731	.0316	.0809	.1199	.8408				
	0 5	-.2272	-.0284	.0644	1.1912					
	0 6	-.5664	-.1521	1.7185						
	0 7	-1.6868	2.6868							
	1 1		.2097	.1159	.1162	.1163	.1162	.1159	.2097	
	1 2		.1626	.1074	.1148	.1214	.1275	.3663		
	1 3		.0799	.0936	.1140	.1321	.5804			
	1 4		-.0768	.0699	.1153	.8916				
	1 5		-.4272	.0218	1.4054					
	1 6		-1.5874	2.5874						
	2 2			.3134	.1243	.1246	.1243	.3134		
	2 3			.2040	.1191	.1330	.5440			
	2 4			-.0527	.1098	.9429				
	2 5			-.9229	1.9229					
	3 3				.4315	.1370	.4315			
	3 4				.0000	1.0000				
10	0 1	.0843	.0921	.0957	.0986	.1011	.1036	.1060	.1085	.2101
	0 2	.0605	.0804	.0898	.0972	.1037	.1099	.1161	.3424	
	0 3	.0244	.0636	.0818	.0962	.1089	.1207	.5045		
	0 4	-.0316	.0383	.0707	.0962	.1185	.7078			
	0 5	-.1240	-.0016	.0549	.0990	.9718				
	0 6	-.2923	-.0709	.0305	1.3327					
	0 7	-.6596	-.2138	1.8734						
	0 8	-1.8634	2.8634							
	1 1		.1884	.1036	.1040	.1041	.1041	.1040	.1036	.1884
	1 2		.1525	.0961	.1013	.1057	.1098	.1138	.3209	
	1 3		.0942	.0846	.0979	.1095	.1204	.4933		
	1 4		-.0043	.0665	.0938	.1179	.7261			
	1 5		-.1866	.0351	.0892	1.0623				
	1 6		-.5877	-.0289	1.6166					
	1 7		-1.9000	2.9000						
	2 2			.2798	.1099	.1103	.1103	.1099	.2798	
	2 3			.2050	.1038	.1122	.1198	.4592		
	2 4			.0606	.0935	.1178	.7281			
	2 5			-.2648	.0735	1.1914				
	2 6			-1.3406	2.3406					
	3 3				.3807	.1193	.1193	.3807		
	3 4				.1871	.1198	.6930			
	3 5				-.4847	1.4847				
	4 4					.5000	.5000			

Source: From Sarhan and Greenberg (1956).

TABLE A5 The Coefficients of the Best Linear Estimates of the Standard Deviation in Censored Samples up to Sample Size 10

n	n_1	n_r	x(1)	x(2)	x(3)	x(4)	x(5)	x(6)	x(7)
3	0	1	-1.1816	1.1816					
4	0	1	-.6971	-.1268	.8239				
	0	2	-1.3654	1.3654					
	1	1		-1.6834	1.6834				
5	0	1	-.5117	-.1668	.0274	.6511			
	0	2	-.7696	-.2121	.9817				
	0	3	-1.4971	1.4971					
	1	1		-1.0101	.0000	1.0101			
	1	2		-2.0201	2.0201				
6	0	1	-.4097	- .1685	-.0406	.0740	.5448		
	0	2	-.5528	-.2091	-.0290	.7909			
	0	3	-.8244	-.2760	1.1004				
	0	4	-1.5988	1.5988					
	1	1		-.7531	-.0829	.0829	.7531		
	1	2		-1.1438	-.0878	1.2317			
	1	3		-2.2717	2.2717				
7	0	1	-.3440	-.1610	-.0681	.0114	.0901	.4716	
	0	2	-.4370	-.1943	-.0718	.0321	.6709		
	0	3	-.5848	-.2428	-.0717	.8994			
	0	4	-.8682	-.3269	1.1951				
	0	5	-1.6812	1.6812					
	1	1		-.6108	-.1061	.0000	.1061	.6108	
	1	2		-.8288	-.1258	.0248	.9298		
	1	3		-1.2483	-.1548	1.4030			
	1	4		-2.4712	2.4712				
	2	2			-1.4176	.0000	1.4176		
	2	3			-2.8352	2.8352			
8	0	1	-.2978	-.1515	-.0796	-.0200	.0364	.0951	.4175
	0	2	-.3638	-.1788	-.0881	-.0132	.0570	.5868	
	0	3	-.4586	-.2156	-.0970	.0002	.7709		
	0	4	-.6110	-.2707	-.1061	.9878			
	0	5	-.9045	-.3690	1.2735				
	0	6	-1.7502	1.7502					
	1	1		-.5184	-.1115	-.0361	.0361	.1115	.5184
	1	2		-.6608	-.1319	-.0318	.0630	.7615	
	1	3		-.8894	-.1605	-.0197	1.0696		
	1	4		-1.3337	-.2086	1.5423			
	1	5		-2.6357	2.6357				
	2	2			-1.0357	-.0674	.0674	1.0357	
	2	3			-1.5661	-.0678	1.6338		
	2	4			-3.1220	3.1220			
	3	3				-3.2784	3.2784		

TABLE A5 (continued)

n	n_1	n_r	x(1)	x(2)	x(3)	x(4)	x(5)	x(6)	x(7)	x(8)	x(9)
9	0	1	-.2633	-.1421	-.0841	-.0370	.0062	.0492	.0954	.3757	
	0	2	-.3129	-.1647	-.0938	-.0364	.0160	.0678	.5239		
	0	3	-.3797	-.1936	-.1048	-.0333	.0317	.6797			
	0	4	-.4766	-.2335	-.1181	-.0256	.8537				
	0	5	-.6330	-.2944	-.1348	1.0622					
	0	6	-.9355	-.4047	1.3402						
	0	7	-1.8092	1.8092							
	1	1		-.4527	-.1107	-.0532	.0000	.0532	.1107	.4527	
	1	2		-.5544	-.1291	-.0563	.0109	.0775	.6514		
	1	3		-.7015	-.1535	-.0578	.0299	.8828			
	1	4		-.9399	-.1896	-.0558	1.1852				
	1	5		-1.4057	-.2534	1.6591					
	1	6		-2.7753	2.7753						
	2	2			-.8817	-.0885	.0000	.0885	.8317		
	2	3			-1.1222	-.1023	.0223	1.2022			
	2	4			-1.6894	-.1227	1.8122				
	2	5			-3.3620	3.3620					
	3	3				-1.8213	.0000	1.8213			
	3	4				-3.6426	3.6426				
10	0	1	-.2364	-.1334	-.0851	-.0465	-.0119	.0215	.0559	.0937	.3423
	0	2	-.2753	-.1523	-.0947	-.0488	-.0077	.0319	.0722	.4746	
	0	3	-.3252	-.1758	-.1058	-.0502	-.0006	.0469	.6107		
	0	4	-.3930	-.2063	-.1192	-.0501	.0111	.7576			
	0	5	-.4919	-.2491	-.1362	-.0472	.9243				
	0	6	-.6520	-.3150	-.1593	1.1263					
	0	7	-.9625	-.4357	1.3981						
	0	8	-1.8608	1.8608							
	1	1		-.4034	-.1074	-.0616	-.0201	.0201	.0616	.1074	.4034
	1	2		-.4803	-.1235	-.0674	-.0166	.0325	.0827	.5726	
	1	3		-.5842	-.1440	-.0734	-.0097	.0514	.7599		
	1	4		-.7359	-.1719	-.0797	.0031	.9844			
	1	5		-.9831	-.2145	-.0859	1.2835				
	1	6		-1.4678	-.2918	1.7595					
	1	7		-2.8960	2.8960						
	2	2			-.7021	-.0947	-.0310	.0310	.0947	.7021	
	2	3			-.8898	-.1101	-.0262	.0549	.9711		
	2	4			-1.1952	-.1318	-.0144	1.3415			
	2	5			-1.7947	-.1688	1.9635				
	2	6			-3.5677	3.5677					
	3	3				-1.2832	-.0559	.0559	1.2832		
	3	4				-1.9791	-.0553	2.0344			
	3	5				-3.9511	3.9511				
	4	4					-4.0761	4.0761			

Source: From Sarhan and Greenberg (1956).

TABLE A6 Variances and Covariances of the Estimates of the Mean (μ) and Standard Deviation (σ) for Censored Samples up to Size 10 from a Normal Population

n	n_1		n_r								
			0	1	2	3	4	5	6	7	8
2	0	$V(\mu^*)$	.50000000								
		$V(\sigma^*)$	.57079633								
		$Cov(\mu^*, \sigma^*)$	0								
3	0	$V(\mu^*)$	.33333333	.44867111							
		$V(\sigma^*)$	.27548197	.63782627							
		$Cov(\mu^*, \sigma^*)$	0	.20443088							
4	0	$V(\mu^*)$	.25000000	.28701097	.51299280						
		$V(\sigma^*)$	.18005021	.30207500	.67303155						
		$Cov(\mu^*, \sigma^*)$	0	.06720309	.35673623						
	1	$V(\mu^*)$		.29819962							
		$V(\sigma^*)$		.70572111							
		$Cov(\mu^*, \sigma^*)$		0							
5	0	$V(\mu^*)$	.20000000	.21772341	.28393206	.61122918					
		$V(\sigma^*)$	.13332124	.19476301	.31808583	.69570833					
		$Cov(\mu^*, \sigma^*)$	0	.03299936	.12335995	.47492042					
	1	$V(\mu^*)$		.22579422	.28683366						
		$V(\sigma^*)$		.32968861	.74060939						
		$Cov(\mu^*, \sigma^*)$		0	.15837417						
6	0	$V(\mu^*)$	.16666667	.17688040	.20682717	.29985334	.71864005				
		$V(\sigma^*)$	.10570264	.14277434	.20436158	.32919818	.71192848				
		$Cov(\mu^*, \sigma^*)$	0	.01945869	.06240445	.17016841	.57052121				
	1	$V(\mu^*)$		.18249856	.20698549	.32960016					
		$V(\sigma^*)$		.21017023	.34598687	.76280252					
		$Cov(\mu^*, \sigma^*)$		0	.05766917	.28373932					
6	2	$V(\mu^*)$			.21474267						
		$V(\sigma^*)$			.77471679						
		$Cov(\mu^*, \sigma^*)$			0						
7	0	$V(\mu^*)$	.14285714	.14942352	.16600214	.20713842	.32479943	.82636223			
		$V(\sigma^*)$	.08749856	.11233100	.14927902	.21136168	.33752616	.72430317			
		$Cov(\mu^*, \sigma^*)$	0	.01276945	.03751914	.08805478	.20989335	.65033968			
	1	$V(\mu^*)$		.15346296	.16680538	.20946061	.39543079				
		$V(\sigma^*)$		.15269771	.22005773	.35717357	.77852463				
		$Cov(\mu^*, \sigma^*)$		0	.02997909	.10645593	.38638223				
	2	$V(\mu^*)$			.17312828	.21044686					
		$V(\sigma^*)$			.36219907	.79619765					
		$Cov(\mu^*, \sigma^*)$			0	.12726434					
8	0	$V(\mu^*)$	.12500000	.12954025	.13988098	.16225105	.21375841	.35409951	.93099884		
		$V(\sigma^*)$	.07461121	.09242321	.11706951	.15419334	.21678723	.34408411	.73416689		
		$Cov(\mu^*, \sigma^*)$	0	.00899283	.02495718	.05377493	.11055562	.24421558	.71859780		
	1	$V(\mu^*)$		.13255044	.14091823	.16234808	.22331839	.4711252[illegible]			
		$V(\sigma^*)$		.11928888	.15943388	.22720134	.36548926	.79042118			
		$Cov(\mu^*, \sigma^*)$		0	.01832825	.05643660	.14825957	.47276076			
	2	$V(\mu^*)$			.14567896	.16269076	.23922424				
		$V(\sigma^*)$			.22999555	.37323003	.81130952				
		$Cov(\mu^*, \sigma^*)$			0	.04936270	.23246854				
	3	$V(\mu^*)$				.16818086					
		$V(\sigma^*)$				.81706103					
		$Cov(\mu^*, \sigma^*)$				0					

TABLE A6 (continued)

n	n_1		n_r								
			0	1	2	3	4	5	6	7	8
9	0	$V(\mu^*)$	.11111111	.11441700	.12139976	.13515042	.16291237	.22407238	.38537601	1.03129529	
		$V(\sigma^*)$	.06501502	.07842444	.09604682	.12074575	.15810000	.22116450	.34943152	.74228259	
		$Cov(\mu^*, \sigma^*)$	0	.00665933	.01775128	.03618024	.06838314	.13048810	.27432806	.77806418	
	1	$V(\mu^*)$		.11673259	.12244352	.13518326	.16466906	.24349912	.55048580		
		$V(\sigma^*)$		.09758705	.12420684	.16448335	.23270051	.37199378	.79983467		
		$Cov(\mu^*, \sigma^*)$		0	.01232979	.03498177	.07983082	.18461869	.54702931		
	2	$V(\mu^*)$			.12605965	.13612535	.16566283	.28513252			
		$V(\sigma^*)$			.16624730	.23712095	.38138044	.82268191			
		$Cov(\mu^*, \sigma^*)$			0	.02670942	.09198622	.32159927			
	3	$V(\mu^*)$				.14097979	.16610128				
		$V(\sigma^*)$				.38407790	.83165694				
		$Cov(\mu^*, \sigma^*)$				0	.10603703				
10	0	$V(\mu^*)$	.10000000	.10250531	.10749284	.11666340	.13359472	.16641008	.23658560	.41735056	1.12690295
		$V(\sigma^*)$	.05759553	.06806087	.08129462	.09891696	.12372658	.16131247	.22479946	.35390665	.74912245
		$Cov(\mu^*, \sigma^*)$	0	.00512045	.01324471	.02595717	.04645253	.08157225	.14831978	.30108776	.83064069
	1	$V(\mu^*)$		.10433167	.10845984	.11682744	.13388367	.17122372	.26722758	.63041997	
		$V(\sigma^*)$		.08241682	.10135027	.12799713	.16847458	.23711263	.37726676	.80752873	
		$Cov(\mu^*, \sigma^*)$		0	.00884085	.02377304	.05004837	.10067393	.21667108	.61197840	
	2	$V(\mu^*)$			.11126628	.11795477	.13388616	.17665047	.34008123		
		$V(\sigma^*)$			.12920079	.17132071	.24257700	.38772575	.83163934		
		$Cov(\mu^*, \sigma^*)$			0	.01678448	.05047738	.12926308	.39861255		
	3	$V(\mu^*)$				.12182093	.13426986	.18655569			
		$V(\sigma^*)$				.24418850	.39205377	.84259431			
		$Cov(\mu^*, \sigma^*)$				0	.04290412	.19638665			
	4	$V(\mu^*)$					.13832644				
		$V(\sigma^*)$					.84582636				
		$Cov(\mu^*, \sigma^*)$					0				

Source: From Sarhan and Greenberg (1956).

APPENDIX B

Programs

In this appendix we shall first refer to some published algorithms for censored data. Then we shall present algorithms of our own for truncated and censored samples.

B.1 PUBLISHED PROGRAMS

Swan (1969b) published an algorithm for computing maximum likelihood estimates for parameters of a normal distribution from grouped and censored data, which is based on the Newton-Raphson method. A description of the algorithm and a short remark concerning the computer program can be found in Swan (1969a) and Swan (1977).

Wolynetz (1979a) presents an algorithm for the same estimation problem as considered by Swan (1969a) but his is based on the EM (expectation-maximization) algorithm (Dempster et al., 1977). A comparison of the Newton-Raphson and the EM algorithm provided by Wolynetz (1979a) reveals that the EM algorithm requires on the average more iterations for convergence than Swan's procedure. The difference in the mean number of iterations increases as the censoring becomes more severe. A relatively high proportion of non-convergence is reported for the Newton-Raphson algorithm while the EM algorithm always converges. This result may be due to the poor starting values used by Wolynetz; i.e., the initial estimates of $\hat{\mu}$ and $\hat{\sigma}$ were $\mu_0 = \Sigma x_i/n_0$, $\sigma_0^2 = \Sigma(x_i - \mu_0)^2/n_0$. As discussed in Chapter 4 there are simplified estimators available which are very close to the MLE. The MML (Tiku, 1967) or the RML (Perrson and Rootzen, 1977), for instance, may be used as initial estimates, when exact ML estimates are required. Using these starting values, the proportion of cases where the Newton-Raphson algorithm does not converge becomes insignificant. The Newton-Raphson algorithm combined with appropriate initial estimates might thus be preferable to the EM algorithm due to its faster convergence.

Both algorithms assume Type I censored samples, but they may also be used for Type II censored samples. The covariance matrix of course has then to be computed separately.

An algorithm for computing maximum likelihood estimates in a linear model from confined and censored normal data was presented by Wolynetz (1979b). His procedure is based on the EM algorithm as well. An alternate program was given by Glasser (1965). Schmee and Hahn (1981) give a computer program of their ILS algorithm for the linear regression model along with a discussion and an example.

An algorithm for the reciprocal of Mills' Ratio was written by Swan (1969a). This program is frequently used as a subroutine in the programs described above.

Some other programs, which must be purchased, are mentioned by Nelson (1982).

```
      SUBROUTINE TRUNCD(MEAN,STD,UR,UL,XQ,S2,IER,EPS)
      DOUBLE PRECISION FG,FH,VG,VH,F,V
      PI=2.506628274631
CC
CCCCCCCCCCCCCCCCCCCCCCCCCCCCCCCCCCCCCCCCCCCCCCCCCCCCCCCCCCCCCCCCCCCCCCCC
CC    NEWTON RAPHSON METHOD TO COMPUTE MLE FOR
CC
CC    DOUBLY TRUNCATED NORMAL POPULATION
CC
CC    OUTPUT PARAMETER
CC
CC    MEAN = MEAN OF UNTRUNCATED POPULATION
CC    STD  = STANDARD DEVIATION OF UNTRUNCATED POPULATION
CC    IER=Ø  CONVERGENCE
CC    IER=1  NO CONVERGENCE IN 6Ø STEPS
CC
CC    INPUT PARAMETER
CC
CC    UR   = RIGHT TRUNCATION POINT
CC    UL   = LEFT TRUNCATION POINT
CC    XQ   = SAMPLE MEAN
CC    S2   = SAMPLE VARIANCE
CC    EPS  = ACCURACY REQUIRED
CC
CC    SUBROUTINES USED
CC
CC    MDNOR FROM IMSL: COMPUTES THE STANDARD NORMAL CDF
CC
CCCCCCCCCCCCCCCCCCCCCCCCCCCCCCCCCCCCCCCCCCCCCCCCCCCCCCCCCCCCCCCCCCCCCCCC
CC
      REAL MEAN
CC
      X=(UR-XQ)/(UR-UL)
      Y=S2/(UR-UL)**2
CC
CC    INITIAL ESTIMATES
CC
      IF(Y.GT.Ø.Ø83) GOTO 2
      H=-Ø.Ø654Ø63+Ø.Ø61463711/SQRT(Ø.Ø833-Y)
      G=X
      A=G/H
      B=(G-1.)/H
CC
CC    NEWTON RAPHSON ITERATION
CC
      DO 1 I=1,6Ø
      ZR=Ø.
      ZL=Ø.
      PDFL=EXP(-B*B/2)/PI
```

```
      PDFR=EXP(-A*A/2)/PI
      CALL MDNOR(A,CDFR)
      CALL MDNOR(B,CDFL)
      ZRH=(A*A*ZR+ZR*(A*ZR-B*ZL))/H
      ZRG=-(A*ZR+ZR*(ZR-ZL))/H
      ZLH=(B*B*ZL+ZL*(A*ZR-B*ZL))/H
      ZLG=-(B*ZL+ZL*(ZR-ZL))/H
      FH=(ZRH-ZLH)*H+ZR-ZL
      FG=(ZRG-ZLG)*H+1.
      VH=2.*H-ZLH*H-ZL
      VG=X-ZLG*H
      F=(ZR-ZL)*H+G-X
      V=H*H-X*X+X*G-ZL*H-Y
      DH=(V*FG-F*VG)/(FH*VG-VH*FG)
      DG=-F/FG-FH/FG*DH
      IF(ABS(DH).LE.EPS.AND.ABS(DG).LE.EPS) GOTO 3
      H=H+DH
      G=G+DG
      IF(H.LT.Ø..OR.G.GT.3..OR.G.LT.-Ø.5) GOTO 2
      A=G/H
      B=(G-1.)/H
1     CONTINUE
2     IER=1
      RETURN
3     MEAN=UR-D*G
      STD=D*H
      IER=Ø
      RETURN
      END
```

```
      SUBROUTINE TRUNCS(U,XQ,S2,MEAN,STD)
CC
CCCCCCCCCCCCCCCCCCCCCCCCCCCCCCCCCCCCCCCCCCCCCCCCCCCCCCCCCCCCCCCCCCCCCCCC
CC
CC    ESTIMATION OF THE MEAN AND STANDARD DEVIATION FOR
CC    SINGLY TRUNCATED SAMPLES
CC
CC    INPUT PARAMETER
CC
CC    XQ  = SAMPLE MEAN
CC    S2  = SAMPLE VARIANCE
CC    U   = TRUNCATION POINT
CC
CC    OUTPUT PARAMETER
CC
CC    MEAN = MEAN OF POPULATION
CC    STD  = STANDARD DEVIATION OF POPULATION
CC
CCCCCCCCCCCCCCCCCCCCCCCCCCCCCCCCCCCCCCCCCCCCCCCCCCCCCCCCCCCCCCCCCCCCCCCC
CC
      REAL MEAN
CC
      A0=-0.00374615
      A1=0.17462558
      A2=-2.87168509
      A3=17.48932655
      A4=-11.91716546
      B1=5.74050101
      B2=-13.53427037
      B3=6.88665552
      C0=0.43780651
      C1=-1.97625969
      C2=3.43594603
      D1=1.00250296
      IF(X-0.57081) 3,3,4
  3   Q=(((A4*X+A3)*X+A2)*X+A1)*X+A0
      P=((B3*X+B2)*X+B1)*X+1.
      GOTO 5
  4   Q=(C2*X+C1)*X+C0
      P=D1*X+1.
  5   MEAN=XQ+Q/P*(U-XQ)
      STD=SQRT(S2+Q/P*(U-XQ)**2)
      RETURN
      END
```

```
      SUBROUTINE VARLIK(ULL,URR,VMY,VMS,COV)
      DOUBLE PRECISION CDFR,CDFL,PDFL,PDFR,ZL,ZR,UL,UR
CC
CCCCCCCCCCCCCCCCCCCCCCCCCCCCCCCCCCCCCCCCCCCCCCCCCCCCCCCCCCCCCCCCCCCCC
CC
CC     ASYMPTOTIC COVARIANCE OF MLE
CC
CC     INPUT PARAMETER
CC
CC     UL = LEFT TRUNCATION POINT
CC     UR = RIGHT TRUNCATION POINT
CC
CC     OUTPUT PARAMETER
CC
CC     VMY = ASYMPTOTIC VARIANCE OF MEAN
CC     VMS = ASYMPTOTIC VARIANCE OF STD
CC     COV = ASYMPTOTIC COVARIANCE
CC
CCCCCCCCCCCCCCCCCCCCCCCCCCCCCCCCCCCCCCCCCCCCCCCCCCCCCCCCCCCCCCCCCCCCC
CC
      UL=ULL
      UR=URR
      PI2=2.506628274631
      PDFR=DEXP(-UR*UR/2)/PI2
      CALL MDNORD(UR,CDFR)
      IF(UL+99.)1,1,2
 1    ZL=0.
      CDFL=0.
      GOTO 3
 2    PDFL=DEXP(-UL*UL/2)/PI2
      CALL MDNORD(UL,CDFL)
      ZL=PDFL/(CDFR-CDFL)
 3    CONTINUE
      ZR=PDFR/(CDFR-CDFL)
C
      E11=1.+UL*ZL-UR*ZR-(ZL-ZR)**2
      E22=-(UR*ZR-UL*ZL)**2+UL*ZL-UR*ZR+2.+UL**3*ZL-UR**3*ZR
      E12=(ZR-ZL)*(1.+UR*ZR-UL*ZL)+UR**2*ZR-UL**2*ZL
      HILF=E11*E22-E12**2
      VMY=E22/HILF
      VMS=E11/HILF
      COV=E12/HILF
      RETURN
      END
```

```
      REAL R(100)
100   FORMAT(29H READ NUMBER OF ITERATIONS M=)
110   FORMAT(33H IF M=0 THEN THE MMLE IS COMPUTED)
101   FORMAT(28H TYPE I OR TYPE II CENSORING)
102   FORMAT(22H READ CONFIDENCE LEVEL)
103   FORMAT(22H READ PERCENTILE      )
104   FORMAT(22H READ N  NL  NR       )
105   FORMAT(22H READ MIN AND MAX     )
106   FORMAT(22H READ DATA            )
107   FORMAT(22H LCI MEAN UCL         ,3F10.6)
108   FORMAT(22H LCI STD  UCL         ,3F10.6)
109   FORMAT(22H LCI PERC UCI         ,3F10.6)
111   FORMAT(22H LOG-TRANSFORMATION ? )
112   FORMAT(22H YES=1   NO=0         )
113   FORMAT(22H MLE OF MEAN AND STD  ,2F10.6)
114   FORMAT(28H BIAS CORRECTED MEAN AND STD)
CC
CCCCCCCCCCCCCCCCCCCCCCCCCCCCCCCCCCCCCCCCCCCCCCCCCCCCCCCCCCCCCCCCCCCCCCCC
CC
CC    PROGRAM FOR COMPUTING THE MLE'S (OR MMLE'S) AND PERCENTILES
CC
CC    CONFIDENCE INTERVALS ARE GIVEN FOR MEAN, STD AND PERCENTILES
CC
CC    BASED ON TYPE I OR TYPE II CENSORED SAMPLES
CC
CC    SUBROUTINE MDNRIS(IMSL) INVERSE NORMAL C.D.F.
CC
CCCCCCCCCCCCCCCCCCCCCCCCCCCCCCCCCCCCCCCCCCCCCCCCCCCCCCCCCCCCCCCCCCCCCCCC
CC
CC    INPUT DATA
CC
      EPS=0.1**5
      PI2=2.506628274631
      WRITE(6,100)
      WRITE(6,110)
      READ(5,*)M
      WRITE(6,101)
      READ(5,*) CENSOR
      WRITE(6,102)
      READ(5,*) P1
      WRITE(6,103)
      READ(5,*) P3
      CALL MDNRIS(P1/2,UP1,IER)
      CALL MDNRIS(P3,UP3,IER)
      WRITE(6,104)
      READ(5,*)N,NL,NR
      WRITE(6,*)N,NL,NR
      WRITE(6,106)
      N0=N-NL-NR
```

```
      NL1=NL+1
      NØ1=N-NR
      READ(5,*)(R(I),I=1,NØ)
      WRITE(6,111)
      WRITE(6,112)
      READ(5,*)ILOG
      IF(ILOG.EQ.Ø) GOTO 11
      DO 1 I=1,NØ
  1   R(I)=ALOG1Ø(R(I))
CC
CC    CALCULATIONS FOR CONFIDENCE INTERVAL AND BIAS CORRECTION
CC
 11   X11=FLOAT(NØ-1)
      PR2=FLOAT(NØ)/FLOAT(N+1)
      IND=Ø
      IF(NL.LT.NR)IND=1
      IF(NL.GT.NR)IND=-1
      NN=N-(1-IND)*NR-(1+IND)*NL
      NNØ=NN-ABS(FLOAT((NR-NL)))
      PR1=FLOAT(NNØ)/FLOAT(NN+1)
      CALL VARLIK(VMY,VSIG,COV,N,NL,NR)
      CHIL=(1.-VSIG/(N*9)-UP1*SQRT(VSIG/(N*9.)))**3
      CHIU=(1.-VSIG/(N*9)+UP1*SQRT(VSIG/(N*9.)))**3
      A1=-COV/(VSIG-N/UP1**2)*2.
      B11=-VMY/(VSIG-N/UP1**2)
      Z1=-A1/2+SQRT(ABS(A1**2/4+B11))
      Z11=Z1+A1
      AH=(VSIG-N/UP1**2)
      A2=-(COV+UP3*VSIG)/AH
      B2=-(VMY+VSIG*UP3**2+2*UP3*COV)/AH
      Z2=-A2+SQRT(ABS(A2**2+B2))
      Z22=Z2+A2*2.
CC
CC    COMPUTING SAMPLE MEAN AND STANDARD DEVIATION
CC
      CALL SORT(R,IRR,NØ,Ø)
      XQ=Ø.
      S2=Ø.
      IF(CENSOR.EQ.2.) GOTO 3
      WRITE(6,1Ø5)
      READ(5,*)XMIN,XMAX
      IF(ILOG.NE.Ø) XMIN=ALOG1Ø(XMIN)
      IF(ILOG.NE.Ø) XMAX=ALOG1Ø(XMAX)
      WRITE(6,*)XMIN,XMAX
      GOTO 4
  3   XMAX=R(NØ)
      XMIN=R(1)
```

```
 4    DO 2 I2=1,N0
      XQ=XQ+R(I2)
      S2=S2+R(I2)**2
 2    CONTINUE
      XQ=XQ/N0
      S2=ABS(S2/N0-XQ**2)
CC
CC    COMPUTING MLE
CC
      CALL MLE01(XQ,S2,XMAX,XMIN,N,NL,NR,M,EPS,IER,XMEAN,STD)
CC
      WRITE(6,113) XMEAN,STD
      WRITE(6,114)
CC
      XMEAN=XMEAN+IND*STD*BMY(PR1)/FLOAT(N+1)
      STD=STD+STD*BSIG(PR2)/FLOAT(N+1)
CC
      XMYL=XMEAN-Z1*STD
      XMYU=XMEAN+Z11*STD
      WRITE(6,107) XMYL,XMEAN,XMYU
      YXU=STD/CHIU
      YXL=STD/CHIL
      WRITE(6,108)YXL,STD,YXU
CC
      PERCEN=XMEAN+UP3*STD
      PERCU=XMEAN+STD*(UP3+Z22)
      PERCL=XMEAN+STD*(UP3-Z2)
      WRITE(6,109)PERCL,PERCEN,PERCU
CC
      STOP
      END
```

```
      SUBROUTINE SORT(X,IRANK,NUMB,ORD)
CC
CCCCCCCCCCCCCCCCCCCCCCCCCCCCCCCCCCCCCCCCCCCCCCCCCCCCCCCCCCCCCCCCCCCCC
CC
CC    SORT  PROGRAM
CC
CCCCCCCCCCCCCCCCCCCCCCCCCCCCCCCCCCCCCCCCCCCCCCCCCCCCCCCCCCCCCCCCCCCCC
CC
      INTEGER IRANK(1),ORD
      REAL X(1),AIM
      IF(NUMB.LE.Ø) RETURN
      IF(ORD) 21,2,21
 21   DO 1 I=1,NUMB
      IRANK(I)=I
 1    CONTINUE
 2    I=1
 3    I=I+I
      IF(I-NUMB) 3,3,31
 31   M=I-1
 4    M=M/2
      IF(M)41,99,41
 41   L=NUMB-M
      DO 6 J=1,L
      I=J+M
      DO 5 II=1,J,M
      AIM=X(I)
      IF(ORD)42,43,42
 42   IND=IRANK(I)
 43   I=I-M
      IF(AIM-X(I)) 44,6,6
 44   K=I+M
      X(K)=X(I)
      X(I)=AIM
      IF(ORD) 45,5,45
 45   IRANK(K)=IRANK(I)
      IRANK(I)=IND
 5    CONTINUE
 6    CONTINUE
      GOTO 4
 99   RETURN
      END
```

```
      SUBROUTINE RMLE1(XXQ,S2,XXMAX,XXMIN,N,NL,NR,MEAN,STD)
      REAL MEAN
      PI2=2.506628274631
CC
CCCCCCCCCCCCCCCCCCCCCCCCCCCCCCCCCCCCCCCCCCCCCCCCCCCCCCCCCCCCCCCCCCCCCCC
CC
CC        RML-ESTIMATES OF THE MEAN AND STD OF A NORMAL
CC        DISTRIBUTION FOR TYPE I OR TYPE II CENSORED SAMPLES
CC
CC        INPUT PARAMETER
CC        XXQ =   SAMPLE MEAN
CC        S2  =   SAMPLE VARIANCE
CC        XXMAX = LARGEST VALUE
CC        XXMIN = SMALLEST VALUE
CC        N   =   SAMPLE SIZE (UNCENSORED)
CC        NL  =   NUMBER OF CENSORED OBSERVATIONS AT THE LEFT
CC        NR  =   NUMBER OF CENSORED OBSERVATIONS AT THE RIGHT
CC
CC
CC        OUTPUT PARAMETER
CC        MEAN = ESTIMATE OF THE MEAN WITH THE RML-METHOD
CC        STD  = ESTIMATE OF STANDARD DEVIATION WITH THE RML-METHOD
CC
CC        SUBROUTINE USED: MDNRIS(IMSL) = INVERSE NORMAL C.D.F.
CC
CCCCCCCCCCCCCCCCCCCCCCCCCCCCCCCCCCCCCCCCCCCCCCCCCCCCCCCCCCCCCCCCCCCCCCCC
CC
CC
      XQ=XXQ
      XMAX=XXMAX
      XMIN=XXMIN
      PR=FLOAT(NR)/FLOAT(N)
      PL=FLOAT(NL)/FLOAT(N)
      IND=0
      IF(PR.EQ.0.)IND=-1
      IF(PL.EQ.0.)IND=1
      IF(IND.NE.0.OR.NL.LE.NR) GOTO 11
      XQ=-XQ
      XMAX=-XXMIN
      XMIN=-XXMAX
      PR=FLOAT(NL)/FLOAT(N)
      PL=FLOAT(NR)/FLOAT(N)
   11 IF(IND.GE.0) GOTO 1
      XQ=-XQ
      XMAX=-XMIN
      PR=PL
      PL=0.
```

```
    1 CONTINUE
      A=Ø.
      B=Ø.
      ZL=Ø.
      ZR=Ø.
      NØ=N-NR-NL
      YQ=XQ-XMAX
      YS=S2+XQ**2-2.*XMAX*XQ+XMAX**2
CC
CC
CC
CC
      CALL MDNRIS (1.-PR,A,IER1)
      ZR=EXP(-A*A/2.)/(PI2*(1.-PR-PL))
CC
      IF(IND.NE.Ø) GOTO 2
      CALL MDNRIS(PL,B,IER1)
      ZL=EXP(-B*B/2.)/(PI2*(1.-PR-PL))
      AELP=YQ*A+ZL*(XMAX-XMIN)
      GOTO 3
    2 AELP=YQ*A
CC
CC
CC
    3 SIG=(AELP+SQRT(ABS(AELP**2+4.*YS)))/2.
      DELT=ZL-ZR
      HELP=B*ZL-A*ZR-ABS(DELT)**2
      MEAN=XQ-SIG*DELT
      IF(IND.EQ.-1) MEAN=-MEAN
      IF(IND.EQ.Ø.AND.NL.GT.NR) MEAN=-MEAN
      SIG=S2-ABS(SIG)**2*HELP
      IF(SIG.LE.S2) SIG=S2
      STD=SQRT(SIG)
      RETURN
      END
```

```
      SUBROUTINE MMLE1(XXQ,S2,XXMAX,XXMIN,N,NL,NR,MEAN,STD)
      REAL MEAN
      PI2=2.50662827463l
CC
CCCCCCCCCCCCCCCCCCCCCCCCCCCCCCCCCCCCCCCCCCCCCCCCCCCCCCCCCCCCCCCC
CC
CC        MML-ESTIMATES OF THE MEAN AND STD OF A NORMAL
CC        DISTRIBUTION FROM TYPE I OR TYPE II CENSORED SAMPLES
CC
CC        INPUT PARAMETER
CC        XXQ =    SAMPLE MEAN
CC        S2  =    SAMPLE VARIANCE
CC        XXMAX = LARGEST VALUE
CC        XXMIN = SMALLEST VALUE
CC        N   =    SAMPLE SIZE (UNCENSORED)
CC        NL  =    NUMBER OF OBSERVATIONS CENSORED AT THE LEFT
CC        NR  =    NUMBER OF OBSERVATIONS CENSORED AT THE RIGHT
CC
CC
CC        OUTPUT PARAMETER
CC        MEAN =    ESTIMATE OF THE MEAN WITH THE MML-METHOD
CC        STD  =    ESTIMATE OF STANDARDDEVIATION WITH THE MML-METHOD
CC
CC        SUBROUTINE USED: MDNRIS(IMSL) = INVERSE NORMAL C.D.F.
CC
CCCCCCCCCCCCCCCCCCCCCCCCCCCCCCCCCCCCCCCCCCCCCCCCCCCCCCCCCCCCCCCC
CC
CC
      XMAX=XXMAX
      XMIN=XXMIN
      XQ=XXQ
      PR=FLOAT(NR)/FLOAT(N)
      PL=FLOAT(NL)/FLOAT(N)
      IND=0
      IF(PR.EQ.0.)IND=-1
      IF(PL.EQ.0.)IND=1
      IF(IND.GE.0)GOTO 1
      XQ=-XQ
      XMAX=-XXMIN
      XMIN=-XXMAX
      PR=PL
    1 A=0.
      B=0.
      BETL=0.
      BETR=0.
      ALFR=0.
      ALFL=0.
      N0=N-NR-NL
```

```
      YQ=(XMAX-XQ)/(XMAX-XMIN)
      YS=S2/(XMAX-XMIN)**2
CC
CC
CC
      CALL MDNRIS (1.-PR,A,IER1)
      BETR=-EXP(-A*A/2.)/PI2*(A-EXP(-A*A/2)/PI2/PR)/PR
      ALFR=EXP(-A*A/2)/PI2/PR-BETR*A
CC
      IF(IND.NE.Ø) GOTO 2
      CALL MDNRIS(PL,B,IER1)
      BETL=-EXP(-B*B/2.)/PI2*(B+EXP(B*B/2)/PI2/PL)/PL
      ALFL=EXP(-B*B/2)/PI2/PL-BETL*B
CC
    2 CONTINUE
CC
      ALFR=ALFR*PR*FLOAT(N)/FLOAT(NØ)
      ALFL=ALFL*PL*FLOAT(N)/FLOAT(NØ)
      BETR=BETR*PR*FLOAT(N)/FLOAT(NØ)
      BETL=BETL*PL*FLOAT(N)/FLOAT(NØ)
CC
      BB=(ALFL+(YQ-BETL)*(ALFR-ALFL)/(1+BETR-BETL))/2
      CC=YS+YQ**2-BETL-(YQ-BETL)**2/(1+BETR-BETL)
      HB=BB*BB+CC
      IF(HB.LE.Ø.) H=SQRT(S2)/(XMAX-XMIN)
      IF(HB.GT.Ø.) H=BB+SQRT(HB)
      G=(YQ-BETL)/(1+BETR-BETL)-(ALFR-ALFL)*H/(1+BETR-BETL)
      MEAN=XMAX-(XMAX-XMIN)*G
      IF(IND.EQ.-1) MEAN=-MEAN
      STD=H*(XMAX-XMIN)
CC
      RETURN
      END
```

```
      FUNCTION BMY(X)
CC
CCCCCCCCCCCCCCCCCCCCCCCCCCCCCCCCCCCCCCCCCCCCCCCCCCCCCCCCCCCCCCCCCCCCC
CC
CC    BIAS CORRECTION FOR THE MEAN
CC    BASED ON SAW'S TABLES
CC
CCCCCCCCCCCCCCCCCCCCCCCCCCCCCCCCCCCCCCCCCCCCCCCCCCCCCCCCCCCCCCCCCCCCC
CC
      BMY=EXP(2.692132-5.438842*X)
      RETURN
      END


      FUNCTION BSIG(X)
CC
CCCCCCCCCCCCCCCCCCCCCCCCCCCCCCCCCCCCCCCCCCCCCCCCCCCCCCCCCCCCCCCCCCCCC
CC
CC    BIAS CORRECTION FOR THE STANDARD DEVIATION
CC    BASED ON SAW'S TABLES
CC
CCCCCCCCCCCCCCCCCCCCCCCCCCCCCCCCCCCCCCCCCCCCCCCCCCCCCCCCCCCCCCCCCCCCC
CC
      BSIG=1./(Ø.312113+Ø.859481*X)**2
      RETURN
      END
```

```
      SUBROUTINE VARLIK(VMY,VSIG,COV,N,NL,NR)
CC
CCCCCCCCCCCCCCCCCCCCCCCCCCCCCCCCCCCCCCCCCCCCCCCCCCCCCCCCCCCCCCCCCCCCCCC
CC
CC    ASYMPTOTIC VARIANCE FACTORS OF MLE FOR TYPE II CENSORED SAMPLES
CC
CC    INPUT PARAMETER
CC    N  =  SAMPLE SIZE
CC    NL =  NUMBER OF OBSERVATIONS CENSORED AT THE LEFT
CC    NR =  NUMBER OF OBSERVATIONS CENSORED AT THE RIGHT
CC
CC    OUTPUT PARAMETERS
CC    VMY =  N * ASYMPTOTIC VARIANCE OF MLE (MEAN)
CC    VSIG = N * ASYMPTOTIC VARIANCE OF MLE (STD)
CC    COV =  N * ASYMPTOTIC COVARIANCE OF MLE (MEAN,STD)
CC
CC
CCCCCCCCCCCCCCCCCCCCCCCCCCCCCCCCCCCCCCCCCCCCCCCCCCCCCCCCCCCCCCCCCCCCCCC
CC
      PI2=2.506628274631
      N0=N-NL-NR
      PR=FLOAT(NR)/FLOAT(N)
      PL=FLOAT(NL)/FLOAT(N)
      QR=FLOAT(NR)/FLOAT(N0)
      QL=FLOAT(NL)/FLOAT(N0)
      IF(PR) 1,1,2
  1   WS=0.
      UR=0.
      GOTO 3
  2   CALL MDNRIS(1.-PR,UR,IER2)
      PDFR=EXP(-UR*UR/2.)/PI2
      WS=PDFR/PR
  3   IF(PL) 4,4,5
  4   W=0.
      UL=0.
      GOTO 6
  5   CALL MDNRIS(PL,UL,IER2)
      PDFL=EXP(-UL*UL/2.)/PI2
      W=PDFL/PL
  6   E11=QL*W*(W+UL)+QR*WS*(WS-UR)+1.
      E22=2.+QL*W*UL*(1.+UL*(W+UL))-QR*WS*UR*(1.-UR*(WS-UR))
      E12=QL*W*(1.+UL*(W+UL))-QR*WS*(1.-UR*(WS-UR))
      HELP=E11*E22-E12*E12
      VMY=E22/(HELP*(1.-PR-PL))
      VSIG=E11/(HELP*(1.-PR-PL))
      COV=ABS(E12/(HELP*(1.-PL-PR)))
      RETURN
      END
```

```
      SUBROUTINE MLE01(XQ,S2,XMAX,XMIN,N,NL,NR,M,EPS,IER,MEAN,STD)
      REAL MEAN
      DOUBLE PRECISION FG,FH,VG,VH,F,V
      PI2=2.506628274631
CCC
CCCCCCCCCCCCCCCCCCCCCCCCCCCCCCCCCCCCCCCCCCCCCCCCCCCCCCCCCCCCCCCCC
CCC
CCC   ML-ESTIMATES OF THE PARAMETERS OF A NORMAL DISTRIBUTION
CCC   FOR TYPE I OR TYPE II CENSORED SAMPLES
CCC   ALGORITHM: NEWTON-RAPHSON WITH MMLE AS STARTING VALUES
CCC              IN THE SINGLY CENSORED CASE ENLOSURE METHOD
CCC              IS USED IF THE N&R-METHOD DOES NOT CONVERGE
CCC
CCC   INPUT PARAMETERS
CCC              XQ   =  SAMPLE MEAN
CCC              S2   =  SAMPLE VARIANCE
CCC              N    =  UNCENSORED SAMPLE SIZE
CCC              NL   =  NUMBER OF CENSORED OBSERVATIONS AT THE LEFT
CCC              NR   =  NUMBER OF CENSORED OBSERVATIONS AT THE RIGHT
CCC              XMIN = SMALLEST VALUE
CCC              XMAX = LARGEST VALUE
CCC              EPS  = ERROR ALLOWED
CCC              M    = MAXIMAL NUMBER OF ITERATIONS
CCC                     IF M=0 THEN THE MML ESTIMATES ARE COMPUTED
CCC     OUTPUT  PARAMETERS
CCC              MEAN = MLE OF MEAN
CCC              STD  = MLE OF THE STANDARD DEVIATION
CCC              IER  = 0 CONVERGENCE
CCC                     1 NO CONVERGENCE
CCC              M= NUMBER OF ITERATIONS
CCC
CCC    SUBROUTINE USED: MDNOR(IMSL)= NORMAL C.D.F.
CCC                     STARTING VALUE: MMLE
CCC
CCCCCCCCCCCCCCCCCCCCCCCCCCCCCCCCCCCCCCCCCCCCCCCCCCCCCCCCCCCCCCCCC
CCC
      IER=0
      N0=N-NR-NL
      PR=FLOAT(NR)/FLOAT(N)
      PL=FLOAT(NL)/FLOAT(N)
      QR=FLOAT(NR)/FLOAT(N0)
      QL=FLOAT(NL)/FLOAT(N0)
CC
CC
      IND=0
      IF(PR.EQ.0.)IND=-1
      IF(PL.EQ.0.)IND=1
      A=0.
      B=0.
```

```
      BETL=0.
      BETR=0.
      ALFR=0.
      ALFL=0.
CC
CC
CC    STARTING VALUES MMLE
CC
CC
      CALL MMLE1(XQ,S2,XMAX,XMIN,N,NL,NR,MEAN,STD)
CC
      IF(M.EQ.0)RETURN
CC
      IF(IND.NE.0) GOTO 1
CC
      D=XMAX-XMIN
      X=(XMAX-XQ)/D
      Y=S2/D**2
      A=(XMAX-MEAN)/STD
      B=(XMIN-MEAN)/STD
      H=1./(A-B)
      G=A/(A-B)
CC
CC
CC    NEWTON-RAPHSON ITERATION FOR DOUBLY CENSORED SAMPLES
CC
      DO 101 I=1,M
      A=G/H
      IF(A-6.) 102,102,100
 102  PDFR=EXP(-A*A/2.)/PI2
      CALL MDNOR(A,CDFR)
      ZR=QR*PDFR/(1.-CDFR)
CC
      B=(G-1.)/H
      IF(B+6.) 100,103,103
 103  PDFL=EXP(-B*B/2.)/PI2
      CALL MDNOR(B,CDFL)
      ZL=QL*PDFL/CDFL
CC
CC    DERIVATIVES
CC
      ZRH=QR*A*ZR*(A-ZR)/H
      ZRG=-QR*ZR*(A-ZR)/H
      ZLH=QL*B*ZL*(B+ZL)/H
      ZLG=-QL*ZL*(B+ZL)/H
      FH=(ZRH-ZLH)*H+ZR-ZL
      FG=(ZRG-ZLG)*H+1.
      VH=2.*H-ZLH*H-ZL
      VG=X-ZLG*H
```

```
      F=(ZR-ZL)*H+G-X
      V=H*H-X*X+X*G-ZL*H-Y
CC
CC    ITERATION STEP
CC
      DH=(V*FG-F*VG)/(FH*VG-VH*FG)
      DG=-F/FG-FH/FG*DH
      H=H+DH
      G=G+DG
CC
CC
      IF(ABS(DH).LT.EPS.AND.ABS(DG).LT.EPS) GOTO 104
CC
CC
 101  CONTINUE
      IER=1
 104  CONTINUE
CC
      MEAN=XMAX-D*G
      STD=D*H
      M=I
CC
      RETURN
CC
   1  CONTINUE
CC
      IF(IND.EQ.1) GOTO 2
      XQ=-XQ
      XMAX=-XMIN
      PR=PL
   2  CONTINUE
CC
      R0=PR/(1.-PR)
      Y=(XMAX-IND*MEAN)/STD
CC
      X=S2/(XMAX-XQ)**2
CC
CC    NEWTON RAPHSON ITERATION FOR SINGLY CENSORED SAMPLES
CC
      DO 201 I=1,10
      PDF=EXP(-Y*Y/2.)/PI2
      CALL MDNOR(-Y,CDF)
      W=R0*PDF/CDF
      WD=W*(PDF/CDF-Y)
      FK=(1.-W*(W+Y))/((W+Y)**2)-X
      FKD=(W*(W+Y)-WD*(Y*(W+Y)+2.)-2.)/((W+Y)**3)
      YD=FK/FKD
      Y=Y-YD
```

```
CC
      IF(ABS(YD).LT.EPS) GOTO 2Ø2
CC
 2Ø1  CONTINUE
CC
CC    IF NEWTON RAPHSON DOES NOT CONVERGE THEN ENCLOSURE METHOD IS USED
CC
CC
      Y=(XMAX-IND*MEAN)/STD
      YO=Y+1.
      YU=Y-1.
      DO 2Ø4 I=1,5Ø
      Y=(YO+YU)/2.
      PDF=EXP(-Y**2/2)/PI2
      CALL MDNOR(-Y,CDF)
      W=RØ*PDF/CDF
      PHI=(1.-W*(W+Y))/(W+Y)**2-X
      IF(PHI.LE.Ø.) YO=Y
      IF(PHI.GT.Ø.) YU=Y
      IF(ABS(YO-YU).LT.EPS) GOTO 2Ø2
 2Ø4  CONTINUE
CC
CC
 2Ø2  PDF=EXP(-Y*Y/2.)/PI2
      CALL MDNOR(-Y,CDF)
      W=RØ*PDF/CDF
      F1=W/(W+Y)
      MEAN=XQ+F1*(XMAX-XQ)
      STD=S2+F1*(XMAX-XQ)**2
      IF(STD.LE.S2)STD=S2
      STD=SQRT(STD)
      IF(IND.EQ.-1) MEAN=-MEAN
      M=I
      RETURN
1ØØ   M=Ø
      RETURN
      END
```

REFERENCES

Dempster, A. P., Laird, N. M., and Rubin, D. B. (1977). Maximum likelihood from incomplete data via the EM algorithm (with discussion), Journal of Royal Statistical Society, Ser. B, 39, 1-38.

Glasser, M. (1965). Regression analysis with censored data, Biometrics 21, 300-307.

IMSL (1975), Library 2 Reference Manual, 5th ed. (November International Mathematical and Statistical Libraries, Inc., 6th Floor, GNB Building, 7500 Bellaire Boulevard, Houston, Texas.

Nelson, F. D. (1981). Applied Life Data Analysis, New York, Wiley.

Persson, T. and Rootzen, H. (1977). Simple and highly efficient estimators for a Type I censored normal sample, Biometrika 64, 123-128.

Schmee, J. and Hahn, G. J. (1979). A simple method for regression analysis with censored data, Technometrics 21, 417-432.

Schmee, J. and Hahn, G. J. (1981a). A computer program for simple linear regression with censored data, Journal of Quality Technology 13, 264-269.

Schmee, J. and Hahn, G. J. (1981b). Correction to "A computer program for simple linear regression with censored data," Journal of Quality Technology 14, 235.

Swan, A. V. (1969a). Computing maximum likelihood estimates for parameters of the normal distribution from grouped and censored data, Applied Statistics 18, 65-69.

Swan, A. V. (1969b). Algorithm AS16, Maximum likelihood estimation from grouped and censored normal data, Applied Statistics 18, 110-114.

Swan, A. V. (1979c). Algorithm AS17, The reciprocal of Mills' ratio, Applied Statistics 18, 115-116.

Swan, A. V. (1977). A remark on Algorithm AS16, Maximum likelihood estimation from grouped and censored normal data, Applied Statistics 26, 122.

Tiku, M. L. (1967). Estimating the mean and standard deviation from a censored normal sample, Biometrika 54, 155-165.

Wolynetz, M. S. (1979a). Maximum likelihood estimation from confined and censored normal data, Algorithm AS138, Applied Statistics 28, 185-195.

Wolynetz, M. S. (1979b). Maximum likelihood estimation in a linear model from confined and censored normal data, Algorithm AS139, Applied Statistics 28, 195-206.

Index